EARTH BIBLE COMMENTARY

Series Editor
Norman C. Habel

Hebrews: An Earth Bible Commentary

A City That Cannot Be Shaken

Jeffrey S. Lamp

t&tclark
LONDON • NEW YORK • OXFORD • NEW DELHI • SYDNEY

T&T CLARK
Bloomsbury Publishing Plc
50 Bedford Square, London, WC1B 3DP, UK
1385 Broadway, New York, NY 10018, USA
29 Earlsfort Terrace, Dublin 2, Ireland

BLOOMSBURY, T&T CLARK and the T&T Clark logo are trademarks of Bloomsbury Publishing Plc

First published in Great Britain 2020
This paperback edition published in 2022

Copyright © Jeffrey S. Lamp, 2020

Jeffrey S. Lamp has asserted his right under the Copyright, Designs and Patents Act, 1988, to be identified as Author of this work.

For legal purposes the Acknowledgments on p. viii constitute an extension of this copyright page.

Scripture quotations are from Common Bible: New Revised Standard Version Bible, copyright © 1989 National Council of the Churches of Christ in the United States of America. Used by permission. All rights reserved worldwide.

Cover design: Charlotte James
Cover image © borchee/istock

All rights reserved. No part of this publication may be reproduced or transmitted in any form or by any means, electronic or mechanical, including photocopying, recording, or any information storage or retrieval system, without prior permission in writing from the publishers.

Bloomsbury Publishing Plc does not have any control over, or responsibility for, any third-party websites referred to or in this book. All internet addresses given in this book were correct at the time of going to press. The author and publisher regret any inconvenience caused if addresses have changed or sites have ceased to exist, but can accept no responsibility for any such changes.

A catalogue record for this book is available from the British Library.

Library of Congress Cataloging-in-Publication Data
Names: Lamp, Jeffrey S., author.
Title: Hebrews: an Earth Bible commentary: a city that cannot be shaken / [Jeffrey S. Lamp].
Description: London; New York: T&T Clark, 2020. | Series: Earth Bible commentary | Includes bibliographical references and index. | Summary: "In this ecological commentary upon the Letter to the Hebrews Jeffrey S. Lamp makes use of the approaches developed in the relatively new field of Ecological Hermeneutics to shed light upon the connection of Hebrews with the Earth. Lamp uses a model of 'suspicion-identification-retrieval' in the light of ecojustice principles in his reading"–Provided by publisher.
Identifiers: LCCN 2019058107 | ISBN 9780567672902 (hardback) | ISBN 9780567672919 (pdf) | ISBN 9780567672926 (epub)
Subjects: LCSH: Bible. Hebrews–Criticism, interpretation, etc. | Human ecology in the Bible. | Human ecology–Religious aspects–Christianity.
Classification: LCC BS2775.52 .L25 2020 | DDC 227/.8706–dc23
LC record available at https://lccn.loc.gov/2019058107

ISBN: HB: 978-0-5676-7290-2
PB: 978-0-5677-0521-1
ePDF: 978-0-5676-7291-9
ePUB: 978-0-5676-7292-6

Typeset by Deanta Global Publishing Services, Chennai, India

To find out more about our authors and books visit www.bloomsbury.com and sign up for our newsletters.

*To
Monica, my wife,
and
Jonah, Kadin, Emily, and Alex,
my grandchildren,
for a better tomorrow*

Contents

Acknowledgments	viii
List of Abbreviations	ix
1 Introduction	1
2 In the Beginning . . . the Son? (Heb. 1:1–2:4)	15
3 Recapitulating Adam (Heb. 2:5-18)	27
4 Building a House on Earth (Heb. 3:1-19)	37
5 Establishing Rest (Heb. 4:1-13)	45
6 A New, Yet Ancient, Priesthood Introduced (Heb. 4:14–6:20)	55
7 Jesus: The New Adamic Priest (Heb. 7:1-28)	67
8 A Logic of the New Order (Heb. 8:1-13)	79
9 A New Order of Worship (Heb. 9:1–10:18)	87
10 Looking for the City Whose Foundations Are in Heaven (Heb. 10:19–11:40)	101
11 The Goal of Creation: The Dwelling Place of God (Heb. 12:1–13:25)	115
12 Conclusion	129
Bibliography	133
Subject Index	140
Ancient Document Index	142
Modern Authors Index	146

Acknowledgments

It takes a lot of people to make a project like this come together. It was about twelve years ago that I first participated with some of the people associated with the Earth Bible Commentary Series. Norman Habel and Peter Trudinger were the first to guide me into the ways of ecological hermeneutics and have been very helpful along the journey. I thank them for their guidance through the years and their encouragement for me to pursue this commentary. I would also like to thank Dominic Mattos, Editorial Director at Bloomsbury—T & T Clark, for his guidance in the writing of this volume, especially for allowing me a couple of extensions on the manuscript submission date due to some health issues I experienced over the past year. I appreciate his flexibility and apologize for any complications it may have caused him and his staff. My colleagues at Oral Roberts University, particularly Wonsuk Ma, Dean of the College of Theology and Ministry, and Samuel Thorpe, Chair of the Undergraduate Department of Theology, have done a wonderful job in creating an atmosphere that encourages scholarly activity, even though Oral Roberts University does not fit into the category of a research institution. I thank them for their encouragement. Finally, I wish to thank my wife Monica, who was quite patient with me during the first summer of her retirement while I was trying to make up for lost time writing the manuscript. She has always been supportive of my activities, and I cannot satisfactorily express my love and appreciation for her. I hope the final product is worthy of all those named here.

Abbreviations

AB	Anchor Bible
BDAG	Walter Bauer, Frederick W. Danker, W. F. Arndt, and F. W. Gingrich, *Greek-English Lexicon of the New Testament and Other Early Christian Literature*, 3rd ed. (Chicago: University of Chicago Press, 2000).
CAFO	Concentrated animal feeding operation
CPT	Centre for Pentecostal Theology
ICC	International Critical Commentary
IPCC	Intergovernmental Panel on Climate Change
IVP	InterVarsity Press
LXX	Septuagint
NIB	New Interpreter's Bible
NIGTC	New International Greek Testament Commentary
NIDNTTE	Moises Silva, ed., *New International Dictionary of New Testament Theology and Exegesis* (Grand Rapids: Zondervan, 2014).
NTL	New Testament Library
WBC	Word Biblical Commentary

1

Introduction

Here We Go Again?

In 2012, I published a collection of ecological readings of Hebrews under the title *The Greening of Hebrews? Ecological Readings in the Letter to the Hebrews*.[1] Yet here we are once again, revisiting this letter from an ecological perspective. The inquisitive reader may rightly ask, "What else is there to say?" A reasonable question. Of course, I might simply respond by saying that the present offering is pursued with a different publisher within a different genre. Whereas the first volume was a series of studies examining selected passages under certain thematic headings, the present volume will consist of a more rigorous ecological commentary on the whole epistle. The scope of the study, then, is perhaps the most obvious difference between the two volumes. But is there something else to say in this regard?

At one level, given the mere passage of time, I would hope that my own learning and reflection in the fields of ecological hermeneutics and ecotheology have advanced such that this volume is not a mere restatement of what was said in the earlier volume. My thinking on the methodology of ecological hermeneutics has resulted in a volume of its own.[2] Many of the considerations in that volume will come to bear on the present study. Whereas the first volume on Hebrews rigorously applied the three criteria of ecological hermeneutics as developed in the Ecological Hermeneutics Section of the Society of Biblical Literature under the guidance of Norman Habel (to be discussed later), this

---

[1] Jeffrey S. Lamp, *The Greening of Hebrews? Ecological Readings in the Letter to the Hebrews* (Eugene: Pickwick, 2012).
[2] Jeffrey S. Lamp, *Reading Green: Tactical Considerations for Reading the Bible Ecologically* (New York: Peter Lang, 2017).

volume will incorporate other tactics and methods in teasing out the ecological message of Hebrews.

So the two primary differentiations between the two volumes are the scope of the present study and the range of methods employed to discern the ecological message of the letter. I trust that readers will see the justification for the present volume as it proceeds.

Challenges in Reading Hebrews Ecologically

Before laying out the methodological approach for this commentary, we must acknowledge several challenges that present themselves as we attempt to address Hebrews ecologically.

Anthropocentric Bias

The first, most obvious, challenge must be the acknowledgment that Hebrews was not written as an ecological treatise. On a surface reading of the epistle, or even on a rigorous exegetical treatment of it, as exegesis is defined and practiced in the present Western academy, the epistle does not have as a definable concern the current well-being of the other-than-human creation. To be sure, as we will point out, there are statements made in the letter that have a bearing on the current ecological crisis facing Earth. Nevertheless, the letter cannot be identified as an ecological treatise in the vein of such works as Rachel Carson's *Silent Spring* or E. O. Wilson's *The Creation*.[3]

Another way to frame this observation is to say that our author and generations of subsequent interpreters have evidenced anthropocentric bias, as seen in the letter. Following Norman Habel's application of this term in ecological hermeneutics, I understand anthropocentric bias to describe a point of view that sees the focus of the biblical writers on human concerns to the neglect of the rest of the Earth community. On the one hand, such a perspective is understandable. After all, it is a document written by a human author in human language to an ostensibly human audience; trees and rivers and giraffes do not read documents such as Hebrews. There was a particular

---

[3] Rachel Carson, *Silent Spring* (Boston: Houghton Mifflin, 1962); E. O. Wilson, *The Creation: An Appeal to Save Life on Earth* (New York: W. W. Norton & Company, 2007).

exigence that gave rise to the author's need to address this audience. Of course, questions abound as to the identity of the author and audience as well as the date, occasion, and purpose of the letter, and even its literary identification as a letter. Still, it is undeniable that regardless of how we assess these matters of critical introduction, Hebrews gives every indication that its primary concern is the well-being of a community of Christian believers regarding their faithful living of the implications of the new covenant established by Jesus Christ. While human concerns dominate the author's message, much is said that has implications for the other-than-human Earth community once we acknowledge our suspicion of anthropocentric bias, work toward points of identification with Earth, and struggle to hear Earth's voice through the biblical text.

Critical Introduction

As noted earlier, there is the matter of ignorance related to issues of critical introduction. Rather than this being an impediment to an ecological reading of Hebrews, this actually frees us to pursue implications for Earth from the text as it presents itself to us. All critical interactions with Hebrews that pursue solutions to matters of authorship, audience, date, occasion, and purpose must live with a degree of provisionality with respect to the conclusions reached on these matters. Moreover, such provisionality extends to readings of the text that derive from these conclusions. Of course, this is an occupational hazard of dealing with many ancient documents. However, the formal anonymity of the author of Hebrews, as well as our lack of knowledge of many other areas of critical introduction, is actually quite liberating. It allows us to approach the text with a certain naivety, to read it on its own terms without appeal to historical reconstructions behind the text. Moreover, we need not be concerned with consonance of thought to an author available in other documents. Our focus can be on reading Hebrews as a stand-alone document from a particular methodological orientation without a compulsion to address matters of critical introduction that not only are largely unknowable, but are ultimately not germane to the methodological orientation we will adopt and describe later.

Dualism

Perhaps the most significant challenge related to an ecological reading of Hebrews has to do with the dualism in the author's argumentation in the letter.

That there is dualism present in the letter is universally acknowledged; the precise nature of the dualism is in question. From early times, the prevailing concern seemed to be the relationship of the dualism in Hebrews to Platonic dualism. While some have seen a close correspondence between Hebrews and Plato, others have argued vehemently against such an identification. The history of debate is long and complex, so here we must be content with broad observations. Platonic dualism posits the existence of a realm of ideas where exists the perfect form of all things, whereas the things on earth are imperfect copies of these ideal forms. So the truly "real" things are in the realm of ideas and may only be perceived via the intellect, while the physical manifestations of these ideas in the world are not ultimately real and are perceived with the senses. This construct has given rise to a rigid physical-nonphysical dualism in which nonphysical reality is exalted at the expense of physical reality. In extreme expressions, the nonphysical—identified with the soul and the spiritual realm—is identified as good, while the physical—identified with the material world—is evil. If Hebrews, then, were to espouse such a dualism, this would have stark ramifications for an ecological reading of the letter. Even given that Hebrews itself is not directly interested in ecological matters, a strong anti-material dualism would certainly direct our ecological reflections in quite another direction. Rather than finding prospects for an implicitly positive, or "green" (to use Norman Habel's label),[4] reading, we would find an implicitly negative, or "grey," reading, and would need to articulate an ecological reading that retrieves Earth's voice in protest against such a reading (more on this later).

The question before us is whether Hebrews presents such a dualism in its argument. Here we must acknowledge that the author of Hebrews uses dualistic language and constructs in the argument of the epistle. Yet it may be reductionistic to ask, what is *the* form of dualism the writer employs? A quick perusal of critical treatments of Hebrews reveals that there are several ways to describe the way the author utilizes dualistic thinking to advance the

---

[4] Norman Habel, *An Inconvenient Text: Is a Green Reading of the Bible Possible?* (Hindmarsh: ATF, 2009). Habel identifies those texts as "green" that depict ecologically positive depictions of God's and humanity's assessment of Earth, while "grey" readings depict ecologically negative assessments. A "grey" reading is not necessarily void of ecologically valuable readings. Whereas a "green" reading would lead to a celebrative reading of a text, a "grey" reading may give rise to Earth's voice raised in protest of and resistance to such a perspective, which has value, for example, in how ecologically minded human beings may act to raise the concerns of Earth to others.

argument of the letter. One way that has much in common with a Platonic conception speaks of a heaven-earth contrast, exemplified in the author's contrast of the imperfect earthly tabernacle that is patterned on the perfect heavenly one (e.g., Heb. 8:5-6; 9:11). However, the author stops far short of asserting that the earthly version of the heavenly ideal is worthless, or more, evil. The author acknowledges at several points that there was some value in what was performed in the earthly tabernacle; it was not perfect, but it was good, and this may be said of all aspects of the former covenant that are described by the author as comparably inferior to what has come to be in the Son. Such an observation has served to temper the assessment of how Platonic the author of Hebrews may have been, at least as discernible in the letter.

A popular way to describe the author's strategy of argumentation is with the heuristic "better than." The letter proceeds, at least through nine and a half chapters, as a series of comparisons that show the new covenant inaugurated through the Son, Jesus Christ, as "better than" the parallel institutions of the old covenant. The strategy and language of comparison betray an array of dualistic conceptions evident in the letter. One such conception is a "covenantal" dualism. The old covenant, represented in such features as its mode of revelation, key figures, priesthood, sanctuary, and sacrifices, has given way to the Son who in his own person and work has instituted a new, better, final covenant. In such a perspective we also see a "historical" dualism at work. The administrations of the two covenants are portrayed as happening in the run of history, with the past giving way to the present. Another dualistic conception portrayed that involves both covenantal and historical dualisms is an "eschatological" dualism. The present age, in which the old covenant has given way to the new covenant on the stage of history, will eventuate with the coming of the new age in which the faithful will find the goal and reward of all their strivings.

This survey of dualism in Hebrews, brief though it may be, does enough to illustrate that dualism is indeed a feature of the author's argumentational arsenal, but that it is overstating the evidence to argue that the author adheres to a rigid Platonic form of dualism that sets spiritual, heavenly realities against material, earthly realities in a good versus evil characterization. We will delve more deeply into the specific instances of dualism in Hebrews as we encounter them. Here we may rest content to observe that the mere presence of a variety of dualistic conceptions is not enough to place the letter into a categorically "grey" reading. As we will see when we turn our attention to methodological considerations for our investigation, the author often provides keys to subvert potentially negative dualistic implications for an ecological reading. To this we now turn our attention.

Methodology

Here we will outline the considerations that will guide our reading of Hebrews. The features of our approach will build consecutively on earlier features to shape, I hope, a coherent reading strategy that teases out ecological implications from the letter.

An Ecological Reading

A heading such as this seems, at first glance, to be mere tautology. Of course an ecological reading of a document would be an ecological reading. But here I wish to draw attention to the specific field of ecological hermeneutics as it has unfolded in the work of the Earth Bible Project and the Exeter Group.

When Lynn White rocked both the Christian and environmentalist communities with his charge that Christianity was largely responsible for the present ecological crisis,[5] several Christian thinkers took up the challenge to defend Christianity from White's accusations.[6] This quest for what may be called the Bible's ecological wisdom produced defenses of the faith that drew from positive depictions of other-than-human creation found in the pages of the Bible. The production of *The Green Bible* (HarperOne, 2008) is exemplary of this approach. Here ecologically positive passages are printed in green ink. However, Norman Habel, among others, quickly recognized that such an approach was naive and tendentious, for the Bible also portrays scenes in which God, and human beings acting ostensibly at the behest of God, do not appear to value the other-than-human world. A more nuanced approach

---

[5] Lynn White, Jr., "The Historical Roots of Our Ecologic Crisis," *Science* 155 (1967): 1203–7.
[6] Rather than rehearse the citations in this space, I direct the reader to my discussion of the development of ecological hermeneutics in *Reading Green*, ch. 1, particularly 8–16, for documentation.

to reading was necessary for a robust ecological hermeneutic. The formation of the Earth Bible Project, working primarily now through the Ecological Hermeneutics Section of the Society of Biblical Literature, developed a series of six ecojustice principles that shaped an approach to reading based on three criteria: suspicion, identification, and retrieval.

By suspicion, Habel means here the anthropocentric bias in the Bible, as we noted earlier. The basic narrative arc of the Bible has as its primary focus the human drama, with other-than-human aspects of the Earth community frequently cast in the role of objects that contribute to the movement of the drama. This, moreover, has manifested in subsequent generations of biblical interpretation, often reinforcing the Bible's anthropocentric bias in the history of biblical interpretation. This is substantially White's critique. In any given instance, once the particular form of anthropocentric bias is identified, ecological interpreters move on to the criterion of identification, where, in the text, points of contact with the other-than-human creation are identified to help foster a sense of empathy with this part of the Earth community. Once the sense of empathy is established, readers employ the criterion of retrieval, where the voice of Earth is teased out of the text. This voice may take many forms: celebration, anger, mourning, protest, prophetic witness, and the like. Not all expressions are necessarily positive, or "green" in Habel's terminology, but they all make a contribution to engagement with the Bible in contemporary discussions of the ecological crises facing humanity and the world.

Another effort at reading the Bible ecologically arose in the first decade of the twenty-first century at the University of Exeter in the United Kingdom. This three-year exploration (2006–2009) engaged scholars from a broad spectrum of academic disciplines to survey the history of ecological engagement with the Bible and to attempt to reconfigure Christian theology and ethics in light of ecologically informed readings of the Bible. The Exeter Group has extended the vision of the Earth Bible Project, it seems, in that it envisions the fruits of its reading of the Bible to contribute to a presentation of Christian theology that is self-consciously ecological in orientation. The Earth Bible Project, on the other hand, in its framing of the six ecojustice principles in largely non-Christian language, seems not to make the move to a reconfiguration of Christian theology, resting content to employ a hermeneutic of suspicion to gain the occasional theological insight.

Our particular ecological approach here draws from both the Earth Bible Project and the Exeter Group. As to the former, we will be conscious of the three criteria of ecological hermeneutics, particularly the criteria of identification and retrieval. My own sense is that the criterion of suspicion may be overdeveloped in many readings. Just because human beings are in view in any given text, this does not always equate with human beings being the *only* subject of value in a text. For example, I have argued that reading through the frame of environmental justice may allow us to read passages that ostensibly deal with poor and marginalized human beings as connected with the suffering of Earth, for often the poor and Earth co-suffer via environmental degradation perpetrated by powerful human beings. Empathy directed toward poor human beings may also facilitate empathy for Earth.[7] However, noting how human beings are engaged in a text may help us move on toward identifying with Earth and allowing Earth's voice to be heard. As to the Exeter Group, I will attempt to engage some of the grand themes of Christian theology in order to articulate them in ecological terms. The letter to the Hebrews is a voice that speaks loudly in doctrinal discussions, contributing to the development of such doctrines as christology (both ontological and economic), soteriology, and eschatology, as we will see. And as we will also see, these doctrines lend themselves quite well to an ecological articulation given the shape of our author's "anthropocentric" presentation.

As I affirmed several times in *Reading Green*, the foundational principle for an ecological reading of the Bible must be the primacy of today's ecological crisis as one approaches the Bible. The ecological crisis has a place of privilege in determining the questions that are posed to the texts. Given that the Bible is not an ecological treatise, its own concerns obviously lie elsewhere. Yet the Bible does indeed make obviously "green" affirmations in its narrative arc, and if properly prompted, may offer implications embedded more deeply in its story that may contribute to contemporary concerns. These prompts must derive from the present crisis facing Earth. This is not to say that every such prompt must be a concrete, specific problem, such as plastics in oceans or global climate change, but that the larger consciousness of the investigation derives from today's ecological crisis. If the Bible were to dictate the parameters of the

---

[7] See *Reading Green*, ch. 2, esp. 39–40, for a treatment of the criteria of ecological hermeneutics in light of environmental justice considerations.

investigation, then the discussion may easily proceed to the grand narrative concerns of its own presentation, drawing our attention away from the largely subordinated ecological statements this story makes along the way. Rather, the ecological crisis serves to focus our attention continually to its own concerns, compelling both the Bible and its readers to stay singular in focus as we readers engage the Bible's grand story. From this ecological perspective, readers will be sensitized to ecological implications of the story perhaps deeply, otherwise imperceptibly, embedded in the narrative. This is not to say that this is the only, or best, way to read the Bible. It merely affirms that this is a valid way to read the Bible for a particular set of concerns. This is how we will proceed in our reading of Hebrews.

A Subversive Reading

The term "subversive" is perhaps troubling to some. However, my use of the term here is not to indicate my desire to engage in textual terrorism of some sort. Rather, it is a term that describes the approach we will take to bridge from our suspicion of the anthropocentrism of Hebrews to our empathetic identification with Earth.

Subversion, in discussions of hermeneutics, often focuses on a pair of dynamics. First, there is the phenomenon that within the pages of the Bible, some writing comes along that subverts the message of an earlier writing. This is writ large in Hebrews. In the author's presentation of the message that the new covenant in the Son is "better than" the old covenant, on several occasions the author takes an aspect of the old covenant scriptures and shows how it was not really speaking as it ostensibly appears, but that in some sense, it points to a reality to which it serves as a precursor. This intra-biblical subversion is quite common throughout the Bible, but especially so in Hebrews. Commentators frequently note that Hebrews serves as an exemplar of early Christian hermeneutics of the Old Testament. One key aspect of this hermeneutic is the author's use of the old covenant scriptures to subvert its own message in favor of the author's presentation of the new covenant.

This form of subversion is quite intentional on the part of the author. Another sense in which the term "subversion" is used is to identify those (probably) unconscious implications in an author's words that open up possibilities for other interpretations and applications of those words. Often this type of subversion is enabled by approaching an author's words from another vantage point. An author, such as the author of Hebrews, has a point to make, and crafts the presentation of that point with language designed to facilitate acceptance of the author's argument. There is an intentionally presented argument offered to support the contentions the author wishes to make. However, if one approaches the author's argument with concerns other than those held by the author, the very words of the author may open up possibilities for understanding that lie outside of the originally intended design of the author.

Of course, these forms of subversion are not entirely discrete from each other. Hebrews is a writing that, seen in canonical perspective, comes along and subverts the message of the old covenant and its scriptures with the message of the new covenant in the Son. However, the way the author of Hebrews accomplishes this is by drawing on the language of the Old Testament to subvert the intention of the Old Testament. For example, take the presentation of the Son's priesthood in Hebrews. As readers, we see that the depiction of the priesthood of Jesus clearly undercuts the validity of the Levitical priesthood described in the Old Testament. So in this broad sense, on this topic, Hebrews subverts the Pentateuchal depiction of priesthood. But how does the author of Hebrews accomplish this? In one concrete instance, the author looks at the prescriptions of the high priest's activities on the Day of Atonement to atone for his own sins (Leviticus 16) as evidence of the weakness of the old covenant priesthood to make true atonement for sins (Heb. 9:6–10:18). Surely this is not the message the writer of the Levitical legislation had in mind, but given the shift in vantage point to the priesthood of the Son, the author is able to glean from the words of Leviticus 16 the grounds for subverting the Levitical priesthood.

As the saying goes, one good turn deserves another. If the author of Hebrews is able to see in the words of the Old Testament the means of subverting those words in favor of the better reality of the Son, it seems that our author has established a precedent that may be employed in reading Hebrews. So we will employ primarily the second type of subversion discussed earlier, where, given our ecological perspective, we will examine the words of our author in order to find a message that subverts the anthropocentrically oriented argument of the author in favor of one that is more affirming of the other-than-human members of the Earth community. One caveat, though. We will not assert, as does the author of Hebrews, that our subversion results in the nullification of the subverted message. The author's subversion of Old Testament language to advance the new covenant in some sense suggests the invalidation of the old covenant. Our use of subversion here will not invalidate the author's argument as far as it goes; it will simply explore the prospect of that argument having implications in an ecological frame of reference.

A Theological Reading

As noted earlier, the Exeter Group has as one of its concerns the reconfiguration of Christian theology in ecological terms. Readings of the Bible from an ecological perspective must, at least in some small part, contribute to such a reconfiguration of theology. An ecological commentary on Hebrews cannot, of course, by itself reconfigure the entirety of Christian theology in this respect. Nevertheless, we can make a modest contribution by adopting a particular theological trajectory to guide our explorations of the letter. The trajectory is suggested by the contours of the author's own argument, and it allows for a reading that subverts the anthropocentric focus of the argument.

The trajectory adopted here proceeds in three movements that depict the Son in his covenantal work. The first movement is that of the Son as the agent of creation. Hebrews 1:1–2:4 establishes the creatorial credentials of the Son, establishing him as the divine revelation par excellence in light of his role in bringing creation into existence and in upholding the created order by his powerful word (1:2-3). In the Son's agency of God's creative work, the Son not only embodies the very nature of God in his incarnate being but also in his creative role mediates a creation that reflects the splendor of God and that will become the inheritance of the Son. The writer of Hebrews begins to describe the new covenant in the Son with creation itself—a creation that has a destiny implicit in the identification of the Son as its heir. In terms of an ecological reading of Hebrews, the proper beginning place, as with the canonical Christian Bible itself, is God's act of creation. Our focus will be on establishing the vocational depiction of human beings within God's creation as a faithful response to the Son through whom, in whom, and for whom all things were made (cf. Col. 1:16). Anticipating the conclusion of this stage of the trajectory, we will argue that the vocational mandate for human beings is to be the priestly co-regents with God to extend God's benevolence into the world that is to be the dwelling place of God.

The second movement of the trajectory encompasses the majority of the epistle, 2:5–10:18, in which the Son's establishment of the new covenant is depicted in terms of the institutions of the old covenant. This large section begins with a reflection on the incarnation of the eternal Son in light of Psalm 8 (Heb. 2:5-18). Given the human vocational mandate and the fact of death and sin in the world, the Son is presented throughout this section of the letter as the Second Adam who comes to reestablish the original vocation of Adam. Here the author's interests in asserting the superiority of the new covenant in the Son over against the old covenant are subverted to show how the mechanics of this new covenant effectively provide a framework for reasserting the original Adamic vocation of human beings in the world. What may seem like an extended hermeneutical foray into reading the Hebrew Scriptures in light of Jesus Christ the Son actually provides the contours of a reading that allows us to glean new covenantal implications for Earth. If, as I will argue, the whole creation was designed to be the dwelling place of God from the beginning, and human beings to serve as God's priestly co-regents in establishing God's benevolence throughout creation, then the new covenant established in the Son provides the contours of understanding how this will be established in the present.

The final movement of the trajectory is found in the actual climax of the epistle, 12:18-29. Here the author develops a contrast between two mountains: Sinai, which represents the old covenant and is characterized as a terrifying sight; and Zion, which is characterized as the New Jerusalem, the kingdom of God. The vision crafted here is one that completes the exhortation of ch. 11, where the author encourages the audience to follow after the patriarchs in their quest to inhabit a new, heavenly city. The faithful, our author asserts, will, and indeed already, inhabit this heavenly city. Though the language seems to suggest that perhaps the earth and the heavens will be shaken and thus removed (vv. 26-27), our investigation will explore possible meanings of this assertion in light of the whole of creation as the inheritance of the eternal Son (1:2). Here we will seek to identify "the created things" (12:27) as things "made with human hands" (a frequent descriptor in Hebrews) that have come to mar creation and hinder its destiny to become the dwelling place of God.

Again, the trajectory here broadly described—creation, Second Adam, New Jerusalem—arguably derives from the interests of the author of Hebrews. What we intend here is to observe in this trajectory points of engagement that inform an ecological reading of Hebrews. To that ecological reading we will turn once we identify one further issue of methodology for our reading of Hebrews.

A Transformational Reading

The commentaries in this series are not commentaries in the sense that the term has come to be expressed in the current world of biblical scholarship. Often those commentaries are highly technical and can span well over 1,000 pages, laden with the seemingly endless multiplication of footnotes. Rather, the series of ecological commentaries to which the current volume belongs is more ideologically and practically oriented. No apology is made for the methodology and orientation of the entries, especially as pertains to the implicit call to action expressed in the desire to retrieve the voice of Earth in reading the biblical text. Once the voice of Earth is heard, readers are confronted with a choice to act on Earth's behalf or to ignore Earth's voice. An ecological reading in this sense is truly a prophetic encounter. It challenges readers to take up the cause of Earth in an age of ecological crises threatening the very existence of life on earth.

In this respect, an ecological commentary more closely represents what constituted biblical commentary in early church history. The worldview and methodologies in those commentaries were thoughtful and rigorous, but the end result was never merely greater understanding of the biblical texts under study. The intended result was a changed reader. In a commentary designed to engage the biblical text from the perspective of Earth, nothing less can be expected. Cold, detached readings are not allowed when the stakes are so high. So here I unapologetically follow in the footsteps of my patristic forebears and aim in this commentary to inculcate in readers a sense of urgency and a call to action to read and act on behalf of Earth. Reading the biblical text in a Christian context should result in a transformation of some sort. Here, we aim for the transformation of mind envisioned by the Apostle Paul in Rom. 12:2.

Terminological Considerations

A word on the choice of terminology used in this commentary is necessary. As the foregoing discussion has illustrated, there is a set of words used that requires distinction: creation, Earth, and earth. Beginning with the last word first, "earth" (lowercase) is used when the planet earth or geological constituency (e.g., dirt, soil) is in view. Uppercase "Earth" is used in the sense adopted by the Earth Bible Project, where the interconnected webs of existence among all things forming the community of life are depicted as a subject with its own integrity. In short, the totality of the biosphere taken as a subject constitutes Earth. "Creation" is used roughly synonymously with "Earth," with the focus on the divine agency through which all things came into existence.

One other thing. Throughout the commentary, I will not use specific gendered pronouns to refer to the author of Hebrews. Given the author's anonymity and some suggestions that the author may have been a woman (e.g., Priscilla), I have chosen to use the term "author" throughout, even though it may seem unnecessarily cumbersome and repetitive at points.

Now with all this being said, let's begin reading Hebrews from the perspective of Earth!

2

In the Beginning . . . the Son? (Heb. 1:1–2:4)

Contents

The opening verses of Hebrews dispense with the typical conventions of opening the letter as typically found in other New Testament epistles. Rather, the argument begins abruptly and dramatically. In a prolonged argument designed to convince readers of the superiority of the new covenant inaugurated by the Son, Jesus Christ, the author begins by establishing the credentials of the Son in the most grandiose terms: the Son is the eternal agent of God's creative work. In the author's argument, this serves to show that the new covenant is established on grounds that far exceed the old covenant. Whereas the old covenant was spoken through prophets and mediated through angels, the new covenant is established through one who is himself the very revelation of God, the one through whom all things come to be, and the one in whom all things hold together. From an ecological vantage point, this perspective, which in reality serves as an interpretation of Gen. 1:1, will function as a basis for unfolding the redemptive work of the Son in terms of a robust understanding of his role as the Second Adam.

Structure

1:1-4: The Son as creator and sustainer of the cosmos
1:5-14: The eternal Son is superior to the angels . . . so what?
2:1-4: A warning more dire than it first seems

Analysis

The Son as Creator and Sustainer of the Cosmos (1:1-4)

Dispensing with the epistolary convention of the salutation, the author immediately introduces the argument of the letter in grand fashion. In what may be termed a historical or eschatological dualism, contrasting the epoch of the prophets of "long ago" with the current revelation that is the Son, the rhetorical strategy of tabbing the Son as "better than" aspects of the old covenant is underway. In what is in Greek one long sentence, vv. 1-4 establish the credentials of the Son in terms of his ontology and his purpose. In an astounding manner, the author lavishes the Son with descriptors that present him as the glorious, cosmic Lord of all creation who also provided for the cleansing of sins in his earthly mission. In so doing, the author has provided us with several foundational considerations that will guide our ecological interpretation of the letter.

If the ancient message was delivered by God through human prophets, the message "in these last days" (v. 2) is delivered in a person, the Son. But this Son is more than simply another prophet. Rather, the Son himself is the revelation of God. Of course, this has provided fodder for christological reflection throughout Christian history, and so has garnered most of the attention paid to the passage. Of particular interest in this respect are two terms that speak to the nature of this Son: ἀπαύγασμα and χαρακτήρ (v. 3). The former term refers to the radiance that emanates from a source, in this instance, God's glory,[1] while the latter term refers to something that is produced to be an exact representation of something else, here the exact representation of God's own essence (Gk. ὑπόστασις).[2] The vocabulary of v. 3 has figured prominently into reflection that has given rise to Trinitarian formulations, particularly in the early centuries of church history. Here we see an anthropocentric bias evident. The passage is mined for how it may satisfy the interests of interpreters.

There are, though, within these verses themselves points of identification that make an ecological reading of the passage, indeed the letter, possible. We may get there by way of posing certain questions of the text. If God has indeed

---

[1] BDAG, 99.
[2] BDAG, 1077-78.

spoken in the person of the Son, what do we see in that person that tells us something of God? What does the Son in fact say about God? If this Son is the radiance of God's glory and the exact representation of God's very being, how does the Son embody these realities?

In this passage, God's activities are depicted in three verbs: speaking, appointing, and making. In the first instance, God spoke in a Son. God's mode of communication is no longer simply verbal; it is personal, embodied, present. Yet God also appointed this Son as the heir of "all things" (πάντα). The final clause of v. 2 both describes this activity and introduces the role of the Son in this activity. God is said to have made (ἐποίησεν) "the worlds" (τοὺς αἰῶνας) through the Son (δι' οὗ) whom God has appointed the heir of all things. It is clearly implied that this Son, who is the eschatological personal embodied presence of God's glory and being, is also the one who prior to the existence of all things was the agent through whom God brought the worlds into being.

It appears that the Son's status as heir of all things is at least in part predicated on his being the agent in the creation of all things. Moreover, our author also attributes the ongoing sustenance of all things (φέρων τὰ πάντα) to the Son's "powerful word" (τῷ ῥήματι τῆς δυνάμεως αὐτοῦ). Here we might see in substance what is affirmed in Col. 1:16-17, where the author asserts that "all things" (τὰ πάντα) were created "in him" (ἐν αὐτῷ) and "through him" (δι' αὐτοῦ) and "for him" (εἰς αὐτόν) and that "all things hold together in him" (τὰ πάντα ἐν αὐτῷ συνέστηκεν).[3] Both the authors of Hebrews and Colossians present what I have called in the earlier monograph "creational christology."[4] The focus in this construct is that a crucial aspect of christology is the involvement of the Son/Christ in creation such that creation itself in some sense bears marks of the Son/Christ and as such is an object of his redemptive work. In Colossians, all things being for him suggests that the destiny of creation is found only in connection with Christ, summed up in the statement that through him all things ultimately are reconciled to him (δι' αὐτοῦ ἀποκαταλλάξαι τὰ πάντα εἰς αὐτον, Col. 1:20). In Hebrews, this finds expression in the appointment of the Son as heir of all things.

These predicates of the Son provide a strong foundation for an ecological reading of Hebrews. If the Son is indeed the personal embodied presence of

---

[3] Lamp, *Reading Green*, 60–61.
[4] Lamp, *The Greening of Hebrews?*, 10–20.

God, a reflection of God's very being, then in the person of the Son we see the God who creates all things through this very Son: a God who lovingly creates all things and then in turn grants them to the Son as his inheritance. Moreover, this Son actively upholds this creation, bridging this role of agent in creation to his inheritance of creation via his personal involvement in the sustenance of creation. That this embodiment of God then inhabits this creation for a time stands as revelation that God cherishes creation enough to invest creation with the divine presence.

Whatever else the Son reveals in these last days in inaugurating the new covenant, which the author of Hebrews will expend great effort in detailing throughout the letter, he embodies the God who is the creator of all things and who sees in creation an inheritance worthy of the Son who embodies God's very glory and being.[5] The Son, in turn, by virtue of his role in creation, sustains this creation with his presence, both terrestrially and cosmically. For in v. 3, the author asserts that once the Son has performed his work in procuring the purification for sins, he ascends and is seated "at the right hand of the Majesty on high." This is language of lordship and will be repeated throughout the remainder of the letter at key junctures. This suggests that the Son's sustenance of all things is achieved through his active rule over creation. The author of Hebrews will draw on this imagery as encouragement for his readers to remain faithful to the new covenant. But if the picture we have drawn here of the significance of the creation of all things in the purview of God is accurate, we will have opportunity to inquire as to what this entails for those living under the active lordship of the Son with respect to his outworking of his sustenance of all things through his subjects.

Moreover, we should not gloss over the reference to the Son's procurement of purification of sins as aimed only toward human beings. I suggest the reference here is anticipatory of later discussions in which purification is significant, and we will leave detailed analysis to those occurrences of the concept. However, at this stage, in a context where the role of the Son is described largely in cosmic terms, it would be odd to assume that the purification of sins is limited only to the human component of creation.

---

[5] Creation as the Son's inheritance was something I neglected to mention in my previous volume, and I am indebted to a reviewer of that book for this insight. See Amy L. B. Peeler, "Review of Jeffrey S. Lamp, *The Greening of Hebrews? Ecological Readings in the Letter to the Hebrews*," *Review of Biblical Literature*, 2014. Available online: http://www.bookreviews.org (accessed July 17, 2019).

In Rom. 8:18-25, Paul describes a creation that groans under the weight of corruption and futility,[6] awaiting the revelation of the children of God at which point the creation will experience its own liberation from the futility to which it was subjected in light of human sin.[7] It may be that in Heb. 1:3, purification of sins has a larger referent in view, with the redemption of human beings from sin entailing the amelioration of the effects of human sin and corruption in the wider created order.[8]

Kenneth Schenck has argued that in the story world of the author of Hebrews, the physical creation serves simply as the temporary and transient stage upon which God works out the drama of the redemption of human beings, though he acknowledges that this passage is perhaps a challenge to this assessment.[9] Our ecological reading of this passage, however, suggests that the physical creation is of much higher value in the divine economy. I suggest that this passage is our author's appropriation of the account of creation in Genesis 1, where God's repeated assessment of creation is that it is good. God reveals Godself in one who physically embodies God's glory and essence in the physical creation and who destines this creation to be the inheritance of the one who was its creatorial agent and who actively sustains it. Moreover, the placement of this passage at the beginning of the argument exercises interpretive control over the portrayal of creation throughout the remainder of the letter. Our ecological reading takes this passage as foundational. If this is the status of creation in God's estimation, this will guide how we read Hebrews ecologically.

---

[6] Against the general trend of reception history, some have seen in this passage not a reference to creation in general awaiting liberation, but of human beings only. For example, Gregory P. Fewster, *Creation Language in Romans 8: A Study in Monosemy*, Linguistic Biblical Studies 8 (Leiden: Brill, 2013); J. Ramsey Michaels, "The Redemption of Our Body: The Riddle of Romans 8:19-22," in *Romans and the People of God: Essays in Honor of Gordon D. Fee on the Occasion of His 65th Birthday*, eds. Sven K. Soderlund and N. T. Wright (Grand Rapids: Eerdmans, 1999), 92-114.

[7] Much discussion focuses on the question as to who precisely subjected creation to futility. Was it Adam? God? Precise identification here is not necessary, for all that need be affirmed is that creation is unable to fulfill its design and purpose due to human disobedience. For discussion, see Sigve K. Tonstad, *The Letter to the Romans: Paul among the Ecologists* (Sheffield: Sheffield Phoenix: 2016), 249-51.

[8] Jon Laansma, "Hidden Stories in Hebrews: Cosmology and Theology," in *A Cloud of Witnesses: The Theology of Hebrews in Its Ancient Contexts*, eds. Richard Bauckham et al. (London: T & T Clark, 2008), 10.

[9] Kenneth L. Schenck, *Cosmology and Eschatology in Hebrews: The Settings of the Sacrifice* (Cambridge: Cambridge University Press, 2007), 139-42.

The Eternal Son Is Superior to the Angels . . . So What? (1:5-14)

At first glance, this passage seems a diversion in the argument. Verse 4 seems to shift dramatically to a statement of the Son's superior status in comparison to angels after the discussion of the Son's role with respect to the created order. Then comes this lengthy catena of scriptural citations that proves the Son's superiority over the angels. It is not until 2:2 that we see the connection. If the Son is a superior revelation to that given through the prophets, the comparison to the angels cements that comparison, for the revelation given through the prophet par excellence, Moses, was actually mediated through angels. If the Son is superior to the angels, then he is superior to the revelation given through the angels. As far as the rhetorical purposes of our author are concerned, the issue is settled. But does this comparison have anything to do with an ecological reading of the epistle?

The answer is yes, but it does so in a way that seems biased against Earth. The crucial passage is 1:10-12, where the author appropriates Ps. 102:25-27 (101:26-28 LXX) to stress the eternality of the Son over against the transience of all things created as demonstration of the Son's superiority over the angels. The passage reads,

> Long ago you laid the foundation of the earth,
>   and the heavens are the work of your hands.
> They will perish, but you endure;
>   they will all wear out like a garment.
> You change them like clothing, and they pass away;
>   but you are the same, and your years have no end.

Schenck sees this as evidence that the material creation is destined for annihilation once its purpose in redemption is served.[10] He links this passage to the discussion in ch. 8, where the verb that is translated "wear out" (παλαιόω) in 1:11 is used in 8:13 to describe the old covenant that is "growing old" and is becoming obsolete. The sense, according to Schenck, is that the creation will face the same fate as the old covenant—it will cease to be once its purpose has been served.

It must be acknowledged here that the passage clearly affirms a subordination of both angels and the physical creation to the Son who endures eternally. As

---

[10] Schenck, *Cosmology and Eschatology in Hebrews*, 122–32.

such, it demonstrates a bias against Earth, or at least a comparative diminution of status. But does this fact merit the conclusion that the author is arguing for a destruction of the physical cosmos? At this point, we may appeal elsewhere in the New Testament to passages that seem to argue for the continuation of the physical order in a state of transformation, such as Rom. 8:18-25 or Revelation 21–22. We may also appeal to our earlier discussion where it seems that the creation is the inheritance of the one who was its agent in its origins. Rather than make these appeals, we may focus on the catena of scriptural quotations to make a counter argument.

Here, the function of the quotation from Ps. 102:25-27 is pertinent. The author is in the midst of a discourse on how the Son is superior to angels. One strand of argumentation is that the Son is indeed just that, a Son, whereas the angels are not. They are servants (Heb. 1:7, 14), and indeed, though entities of a different order than human beings, they are nevertheless creatures. Given the close connection of the angels to the physical creation by virtue of the scriptural citations, we are justified in asking whether the angels, being creatures, are also destined for annihilation. The author gives no hint that the angels, by virtue of their status as inferior to the Son, are thus fated. Given this, we may wonder if it is truly the intention of the author here to argue for the destruction of the cosmos. Or is the passage quoted for the purpose to show that the Son's superiority over the angels is due to the Son's eternality? All created things are by definition not eternal. Schenck may be pushing the details of the quoted passage beyond its intention.

This assertion is supported by another quotation in this catena. In v. 7, the author cites Ps. 104[103 LXX]:4, though altering the LXX's language to emphasize the transitory nature of angels as "winds" and "flames of fire."[11] The important observation here is that the author cites a psalm that is, according to William Brown, a "panoramic sweep of creation from the theological and the cosmological to the ecological and the biological, all bracketed by the doxological."[12] The psalm celebrates the creation from the perspective of the God who created it and rejoices in it. God is depicted as the one who both created and continually sustains the creatures of the world, lovingly

---

[11] Lamp, *Greening of Hebrews?*, 109–11.
[12] William P. Brown, *The Seven Pillars of Creation: The Bible, Science, and the Ecology of Wisdom* (Oxford: Oxford University Press, 2010), 145.

providing for their needs, human and other-than-human alike. This mirrors the ascriptions of the Son in Heb. 1:2-3. Yet this is not the whole of the psalm. In v. 29, the psalmist soberly notes that God may indeed withhold the life-giving Spirit that keeps all creatures thriving. This seems a harsh intrusion into the exultation that fills the psalm to this point. With v. 30, the tone turns more to prayer, voicing the hope that God will fill the earth once again with divine glory, with the final verse (v. 35) of the psalm making one final petition: that God will remove sinners from the earth. The implication is that something has gone wrong with God's beautiful creation and this needs rectification.

I suggest that the full effect of this psalm is squarely in the mind of the author of Hebrews, and that it tempers the details of the citation from Ps. 102:25-27 that follows. The author does not come right out and say this, however. That will come much later in the epistle. Psalm 104 lingers in shadows at this point. The words of Psalm 102 stand explicitly, though not in the way that Schenck suggests. Those words establish the eternal superiority of the Son over angels. The interplay of the two psalms at this point is mere suggestion of the author's ultimate depiction that will come in ch. 12. In 1:3, the Son sits as Lord over all things at the right hand of God after effecting the purification of sins. This is the initial realization of the psalmist's petition in Ps. 104:35. Sin, in all its ramifications for both human beings and the whole of creation, has been remedied, and this will spread to engulf all things. Indeed, the created order is temporal, but that which will be rolled up as a cloak and changed as a garment (v. 12) is the corruption that has infected God's good creation, freeing creation to experience its goodness as the inheritance of the Son. At this point in the letter, though, this is anticipation and requires much more discussion before it can attain to assertion.

A Warning More Dire than It First Seems (2:1-4)

At the beginning of ch. 2, the author offers the first of many warning passages, apparently in response to some exigence in the community of readers, to hold fast to that which they have heard and to which they have pledged fealty. Here we see the reason for the author's extended discourse concerning the angels; the first message was declared through angels to Moses, the exemplar among the prophets of old, and is thus inferior to the message that is the Son. Of course, both the author and subsequent generations of interpreters have understood this as a warning to remain faithful to the covenantal obligations designed to bring the blessings of salvation to the people. But in light of our ecological reading of the preceding chapter, how might we view this warning, and indeed the other warnings, ecologically?

In Heb. 1:1-4, we focused on considerations that directed our attention to the cosmic dimensions in the description of the Son and to how they function foundationally to establish the framework for an ecological reading of Hebrews. The Son is the agent and sustainer of creation, the one whose redemptive work extends beyond human beings to "all things," and the whole of creation is given to him as an inheritance. As revelation, the Son embodies the God who has thus brought all things into being. Though the opening paragraph of ch. 2 focuses on the spoken message that originated with the Lord and was then communicated through those who heard him to the author's circle, the issue in 1:1-4 is the fact that the revelation of God in these last days is in the person of the Son. Of course, this would entail whatever the Son said and did, but that must be understood in connection with the fact that this Son in his person is the reflection of God's glory and the exact representation of God's very nature. And this reveals a God deeply connected with creation. Indeed, the people must heed the spoken message that brings salvation. But it would seem incumbent upon them to pay heed to the totality of that revelation.

In this light, neglecting "so great a salvation" (2:3) may encompass the salvation that is due all things, not just human beings. Ecological readers would insist that the covenant inaugurated through the Son as he is depicted in 1:1-4 would be a covenant that involves the creation that is given the Son as his inheritance. If the Son sits enthroned in heaven, exercising his lordship over the created order, he seems to do so through adherents to his new covenant. The author notes that the message spoken to the audience has been amply confirmed by miraculous occurrences and gifts of the Holy Spirit (2:4). This reference impinges on our ecological reading in two ways. First, it demonstrates the involvement of the risen and enthroned Lord in the earth. In Acts, prior to his ascension to heaven, Jesus tells his disciples that they must await baptism with the Holy Spirit (1:5) so that they may be endued with power to be his witnesses (1:8). In Acts 2, the Holy Spirit comes upon them in a "sound like the rush of a violent wind" (2:2) and gives them "tongues as of fire" (2:3). In Heb. 1:7, the angels were likened to wind and fire; in Acts the Holy Spirit is thus characterized in a demonstration of the power given to the disciples.[13] If the angels are servants of God to assist those who will receive salvation (Heb. 1:14), in Acts the disciples of Jesus become the Spirit-empowered servants of the enthroned Lord in the earth, continuing the mission of the Spirit-empowered Jesus. In the Holy Spirit poured out on his people, the enthroned Lord is present working through them in his inheritance to bring all things to their true destiny. This leads to the second area of significance. If the Spirit is demonstrating the veracity of the message through acts of power and gifts of the Spirit, these should, in some way, show continuity with those performed by Jesus during his earthly ministry. We would do well to note that among the miracles of Jesus was the calming of the storm (Mt. 8:23-27//Mk 4:35-41//Lk. 8:22-25), a miracle that Michael Trainor argues signifies Jesus's healing of the other-than-human created order.[14] Continued acts of Spirit-empowered compassion on the other-than-human creation would seem to be in order.

The warning passages in Hebrews, whatever their specific purpose in the author's argument, each in their own way may extend the anthropocentric emphasis in the letter into an ecological frame by focusing on the role of "all things" as the Son's inheritance and thus as an object of the Son's redemptive ministry. This places all creation within the framework of the superior covenant instituted by the Son. As we did here, we need simply to identify how specifically each warning passage focuses attention on Earth's context within the new covenant.

Hearing Earth

As noted in the introduction to this volume, an ecological reading privileges the ecological crises facing today's world as we plumb the Bible for its contributions to addressing these crises. That there are multiple such crises facing our world is patently obvious to the casual observer. As we conclude each chapter of this commentary, we will seek to hear the voice of Earth, seeing Earth as a subject that experiences the consequences of environmental degradation.

---

[13] Pentecostal theologian A. J. Swoboda highlights the use of creational language to describe the activity of the Holy Spirit. See his *Tongues and Trees: Toward a Pentecostal Ecological Theology*, Journal of Pentecostal Theology Series 40 (Blandford Forum: Deo, 2013), 91.

[14] Michael Trainor, *About Earth's Child: An Ecological Listening to the Gospel of Luke* (Sheffield: Sheffield Phoenix Press, 2012), 153–56; cf. Lamp, *Reading Green*, 128–29.

How might we hear Earth in this section of Hebrews? Earth here will draw our attention to three points. First, Earth will exhort us to remove the blinders from our narrow dogmatic interests as we read this section of Hebrews. As we noted earlier, the opening paragraph of Hebrews serves as fodder in the theological debates prominent in church history, particularly those related to christology and Trinitarian reflection. Once read, the ascription of creation to the agency of the Son loses sight of that which the Son brings into being and focuses on what this assertion says about the Son, both on the part of the author and subsequent interpreters. At this point, Earth urges us to enlarge our focus by recalling that in this paragraph the author also mentions that all things are given the Son as an inheritance. This suggests that perhaps the Son's purification of sins may entail an effect that extends beyond just human benefit. And if the Son sits at God's right hand in heaven, enthroned as the Lord over all things, we should see that indeed the Son's realm extends beyond human interests alone. Earth, in effect, simply asks that we keep the whole paragraph in view, not just those elements that satisfy our own agendas.

Second, Earth will encourage us to extend our vision of what the created order represents in God's dealings with human beings. We noted in our discussion that Heb. 1:2-3 falls into what I have labeled "creational christology." Elsewhere I have argued that the passages in this category (Jn 1:3, 10; 1 Cor. 8:6; Col. 1:16-17; Heb. 1:2-3) are the culmination of a trajectory that spans both testaments.[15] In the Genesis creation accounts, God is said to have created the heavens and earth. In Jewish wisdom, it is first said that God by wisdom created the earth (Prov. 3:19). Later in Proverbs the figure of Wisdom is personified and is depicted as God's agent in creation (8:22-31). In Wis. 7:22–8:1, Wisdom is more highly personified in feminine terms and her role in creation is depicted in highly elaborate terms. Wisdom was the "fashioner of all things" (v. 22); she also pervades and penetrates "all things" (v. 23), renews "all things" (v. 27), and orders "all things" well (8:1). Moreover, she is "a reflection of eternal light/a spotless mirror of the working of God/and an image of [God's] goodness" (7:26). She inhabits the souls of holy persons and makes them friends of God (v. 27), and such persons are characterized as

---

[15] Lamp, *Reading Green*, 57–62. For further discussion, see Jeffrey S. Lamp, *First Corinthians 1–4 in Light of Jewish Wisdom Traditions: Christ, Wisdom, and Spirituality* (Lampeter: Edwin Mellen, 2000), 53–69, 119–29.

those who live with Wisdom (v. 28). The similarity of this depiction to that in Heb. 1:2-3 is striking. In the "creational christology" passages, the final revelation is that it is the Son (or Christ) who plays this role with respect to creation. "Creational christology," when informed by the predicates given wisdom in Jewish wisdom reflection, suggests not only that the Son is active in the creation and sustenance of all things but also that there is something of the Son present in creation that may be discovered by beholding creation. Earth challenges us to see the whole of creation as an icon that points us to the God who created all things through the Son.[16] If the Son is the embodiment of God's glory and very being, then to see creation rightly is to see, if only indirectly, the Son as that revelation of God.

Third, Earth's voice will highlight its own prophetic character. In Heb. 2:1-4, the author's first warning to the audience occurs. In it, the author warns the readers to hold fast to the salvation that is found in the new covenant of the Son in light of the consequences that befall those who reject it. If the expansive vision of the first two points of this discussion of Earth's voice is indeed part of the new covenant salvation inaugurated by the Son, then the warning not to reject this covenant is Earth's assertion of its place in the Son's saving work.

Before any specific ways in which to participate in Earth's healing may be addressed, Earth here calls readers to adopt the proper frame of reference for any such activity on Earth's behalf. One must see the relationship between the Son and Earth in order to understand the foundation for responding to the ecological crises that threaten human and other-than-human existence. The next phase of our discussion moves from consideration of the cosmic portrayal of the Son to the incarnational portrayal, a crucial step in an ecological reading of Hebrews.

---

[16] See John Chryssavgis, "Icons, Liturgy, Saints: Ecological Insights from Orthodox Spirituality," *International Review of Mission* 99 (2010): 182–84; Daniel Munteanu, "Cosmic Liturgy: The Theological Dignity of Creation as a Basis of an Orthodox Ecotheology," *International Journal of Public Theology* 4 (2010): 333.

Recapitulating Adam (Heb. 2:5-18)

Contents

If Hebrews began with the description of the Son in cosmic terms, with this section, the depiction comes crashing down to earth, so to speak. The focus is quite terrestrial, laying out the conceptual framework for the Son's incarnation. The overriding consideration here for the author is establishing the grounds for the Son's salvific work in terms reminiscent of several early church fathers: the Son becomes like we are so that we might become like he is. The point is that he must become human so that he may function as a sympathetic high priest on our behalf. From an ecological perspective, the Son becomes like we are so that he might fulfill the vocation of the first Adam to be God's priestly co-regents in creation (cf. Gen. 1:26-28) and in turn pass that vocation back to those who receive the salvation offered in the new covenant.

Structure

2:5-9: What was the first Adam supposed to be?
2:10-18: What has the Second Adam accomplished?

Analysis

What Was the First Adam Supposed to Be? (2:5-9)

In Heb. 2:5, the author shifts our attention somewhat dramatically to more terrestrial concerns. If Hebrews 1 focused on what classical theology has termed the divinity of the Son, Hebrews 2 shifts more dramatically toward the significance of his humanity. Verse 5 begins with a statement that is transitional. Here, the author states that the "coming world" that is the topic under consideration was not subjected to angels. The mention of angels allows the author to do two things. First, it allows for a transition to the quotation of Ps. 8:5-7, which is thematic for the argument at this point. Second, it affirms that the "coming world," that which was inaugurated by the Son, is firmly grounded in the story of the first creation and the place of human beings in it.

The psalm as quoted in Heb. 2:6-8 reads as follows:

What are human beings that you are mindful of them,
>or mortals, that you care for them?
You have made them for a little while lower than the angels;
>you have crowned them with glory and honor,
>subjecting all things under their feet.

The quotation here draws from Septuagintal language, as seen in the reference to "angels" in Heb. 2:7. The choice of the LXX here provides for a smooth transition to the introduction of human beings into the argument, especially as that pertains to the Son. But in Hebrew, the word translated ἀγγέλους is אלהים (*Elohim*), which is frequently understood as an appellation for God. The Septuagintal reading serves the purpose of the author of Hebrews quite well; the original Hebrew reading opens possibilities for an ecological reading. This occurs through a subversion of our author's quotation by appeal to the Hebrew of the original.

The psalm evokes recollection of Gen. 1:26-28, where the creation of human beings is described from God's perspective. God is said to have created human beings "in the image of God" (Heb. בצלם אלהים; LXX κατ' εἰκόνα θεοῦ, Gen. 1:27). Genesis 1:27 and Ps. 8:6 are linked by the word אלהים. Psalm 8:5-7 is a hymnic meditation on the place of human beings in God's created order (cf. vv. 3-4). This is described both in Genesis and the psalm as a place of dominion over the other living creatures. In Genesis, this is affirmed by repetition of this role in 1:26 and 28. In Ps. 8:7, the scope of human dominion extends more generally to "all" (Heb. כל; LXX πάντα) that God has made. In each case, what is in view is the place of human beings in God's order cast from the perspective of God's original creative design. Human beings are to have dominion over creation.

Norman Habel sees in Gen. 1:26-28 the seeds for human exploitation of Earth.[1] Indeed, this understanding of the passage is substantively behind Lynn White's famous article that attributes responsibility for the current ecological crisis to Christianity.[2] It seems obvious that this passage has been appropriated by parties both Christian and otherwise to justify unbridled exploitation of Earth's resources for human thriving in the world.[3] Such an interpretation needs to be challenged, and has been by many since the publication of White's article.[4] But viewed from within its place in the canonical structure of the Bible, Gen. 1:26-28 actually provides a corrective to interpretations that justify maltreatment of Earth for human ends.[5]

Richard Middleton's important study on image language in the Ancient Near East demonstrates that images were erected in distant parts of a monarch's realm to remind inhabitants of the presence of the monarch in that place despite the monarch's physical absence.[6] Seen in this light, human beings as the image of God are commissioned to extend the rule of God into all parts of God's very good (Gen. 1:31) creation. Early in Christian history, many patristic writers saw this in terms of the priesthood of human beings in the world. Writers such as Maximus the Confessor understood human beings, created uniquely in the image of God as both heavenly and earthly creatures, as a "microcosmos" in which exists the intersection of the celestial and terrestrial, to have a ministry of serving creation on behalf of God, extending God's benevolence into the world, helping creation to attain to its divinely ordained destiny to become

---

[1] Norman C. Habel (*The Birth, the Curse and the Greening of Earth: An Ecological Reading of Genesis 1–11* [Sheffield: Sheffield Phoenix, 2011], 35–40) is representative of those who see in the verbs found in Gen. 1:28, רדה ("rule," "have dominion") and כבש ("subdue"), rather violent actions elsewhere in the Hebrew Bible, and indeed in the whole scene, a framework that implies an anthropocentrism in which there exists a hierarchical structure that puts human beings over the rest of creation, with the verbs supplying the nature of the relationship, which is exploitive.

[2] White, "The Historical Roots of Our Ecologic Crisis," 1203–7.

[3] Lamp, *Reading Green*, 8–9.

[4] One of the very first substantive responses to White was the volume by Francis A. Schaeffer, *Pollution and the Death of Man: The Christian View of Ecology* (Wheaton: Tyndale House, 1970). The number of responses through the years is legion, but perhaps Richard Bauckham's assessment is best: White raises a number of important considerations, and there is some Christian culpability for the present crisis, but White's treatment is truncated and reductionistic because it does not address non-Christian influences on Christianity during the medieval period and it ignores those strands of tradition that put forth resources for a robustly positive model for the relationship of human beings to the rest of nature. See his book, *Living with Other Creatures: Green Exegesis and Theology* (Waco: Baylor University Press, 2011), 15–19.

[5] Terence Fretheim (*The Book of Genesis*, NIB [Nashville: Abingdon, 1994], 346) represents those who see the scene and the verbs in a more positive, nurturing sense.

[6] Richard Middleton, *The Liberating Image: The* Imago Dei *in Genesis 1* (Grand Rapids: Brazos, 2005), ch. 3.

God's dwelling place.[7] We will say more about creation as God's dwelling place later. At this point, it is incumbent to understand that the human role in creation, as per God's original design, is to function as priestly co-regents with God to extend God's benevolent rule throughout all creation. This is essentially an extension of God's command to Adam to "tend and protect"[8] the Garden of Eden to the whole world (Gen. 2:15).

The writer of Hebrews, however, is quick to point out that the ideal picture presented in Psalm 8 is far from realized in the present. While God left "nothing outside their control" (Heb. 2:8), it is clear that at this point, something has gone horribly wrong. Unstated yet implicit is the narrative of Genesis 3, where human beings functionally abdicated their role as priestly co-regents with God. If one dispenses with later interpretations of the serpent in the Garden as Satan or the like, and rather views the serpent simply as a serpent,[9] then what is on display here is human beings, rather than exercising dominion over the serpent, capitulating to the serpent, inverting God's creational design. Human beings were no longer the priestly intermediaries between God and the other-than-human order, but were now subservient to the other-than-human order, a fate sealed by God's decree that human beings would struggle to find survival in the world (Gen. 3:17-19). It is at this point that Habel's and White's critiques become valid, at least in terms of canonical and narrative structure and subsequent appropriation of Gen. 1:26-28.

In Heb. 2:9, the author introduces what will be the crux of our ecological interpretation of the letter: "but we do see Jesus." It is noteworthy that for the first time in the letter, the author is focusing attention to the human being Jesus and not the eternal Son of the previous chapter. This Jesus was "for a little while made lower than the angels." This is a clear reference to the eternal Son's incarnation as a human being,[10] for becoming lower than the angels is

---

[7] See Radu Bordeianu, "Maximus and Ecology: The Relevance of Maximus the Confessor's Theology of Creation for the Present Ecological Crisis," *The Downside Review* 127 (2009): 111–14; Andrew Louth, "Man and Cosmos in St. Maximus the Confessor," in *Toward an Ecology of Transfiguration: Orthodox Christian Perspectives on Environment, Nature, and Creation*, eds. John Chryssavgis and Bruce V. Foltz (New York: Fordham University Press, 2013), 59–71; cf., Maximus, *Ad Thalassium*, 51. See also Gregory of Nyssa, *On the Making of Man*, 2–5.

[8] Sandra Richter, "Environmental Law in Deuteronomy: One Lens on a Biblical Theology for Creation Care," *Bulletin for Biblical Research* 20 (2010): 376.

[9] This is the position espoused by Habel, *The Birth, the Curse and the Greening of Earth*, 57, and Fretheim, *Genesis*, 359–60.

[10] This verse is important to the Eastern Orthodox understanding of the mission of the Son and is celebrated liturgically during the Lenten season and Holy Week.

an attribute of human beings in the quotation of Psalm 8 a few verses earlier. Hebrews 2:9 is a brief encapsulation of the career of the Son. As the one who effected the creation of the world, the Son enters the world as the human being Jesus, procures salvation through the suffering of death, which recalls achieving the purification for sins in 1:3, and resumes his place of glory and honor. How does this verse, then, become the crux of our ecological interpretation of Hebrews?

Here we must look again at the creation narratives of Genesis. In Gen. 2:7, God is said to have formed the human being, which in Hebrew is אדם (*adam*), from the עפר (dust, soil)[11] of the אדמה (ground). Note the lexical connection אדמה/אדם. The physical constitution of the human being derives from Earth, as do the physical constitutions of the animals (Gen. 2:19).[12] Thus human beings are depicted as closely intertwined with Earth. When the eternal Son becomes a human being, he assumes the physical constitution of the human being, which itself derives from Earth.[13] So Jesus, then, assumes not simply the stuff of human beings but also the stuff of Earth. This casts a new light on the patristic dictum that what the Son assumes, he redeems.[14] When Jesus "taste[s] death for everyone" (Heb. 2:9), he tastes death not only on behalf of human beings but also on behalf of Earth.

As we will see, throughout the letter the author will assert that the Son became like us so that he might procure salvation for us. But the Son has also become like Earth, so what he does to procure salvation also extends to Earth. This undercuts the obvious anthropocentric bias toward human beings in the letter. The author cannot be blamed for such an anthropocentric bias; the interests of the letter lie in this sphere. But we can see in the author's own exposition implications for Earth, once we take into account the broader sphere of traditions that lie behind what we see in the letter.

As noted earlier, the current human dominion over Earth is not manifest, at least in the way envisioned in the Genesis creation narratives. While the author notes that we do not currently see the dominion of human beings exercised in the world, we nevertheless see Jesus. The construction in v. 9, with the

---

[11] The Contemporary English Version translates the Hebrew noun as "soil," thus strengthening the connection of human beings to Earth.
[12] Lamp, *The Greening of Hebrews?*, 31–32.
[13] I discuss this concept in terms of the Johannine Prologue (Jn 1:14) in *Reading Green*, 65–67.
[14] Gregory of Nazianzus, in his "Critique of Apollinarius and Apollinarianism," *Epistle 101*.

adversative conjunction δέ, implies that Jesus must, in some way, be the one in whom the picture in the quoted psalm will come to pass. We have noted that in the incarnation of the eternal Son in the person of the human being Jesus a foundation for seeing Earth within the purview of his redemptive mission. But we now see that there is also a sense in which in Jesus, the vocational mandate given to Adam, as reflected in Psalm 8, is recovered. Jesus is in effect the Second Adam, the one in whom the vocational mandate given to the original Adam is realized. The ecological significance of this assertion is profound in two ways. First, this suggests that the original plan of God has not been jettisoned. It will be through human beings that God's benevolent rule in the world is realized. It is not that Jesus is a sort of "Plan B" to salvage something out of the original plan, seen largely in terms of the salvation of human beings. Rather, God has ensured that the redemption of the cosmos will entail putting creation back on track, or to use N. T. Wright's language, "putting the world to rights."[15] And this must be accomplished through human beings. "But we see Jesus" is the affirmation that the Second Adam will establish the loving dominion over creation that was intended for the first Adam.

Second, whatever else Jesus does to achieve salvation, as explicated throughout the letter, it must include the sense that Jesus's newly inaugurated covenant entails the recovery of the Adamic vocation for all human beings to serve and protect Earth. This suggests that prior to the full eschatological realization of salvation, participants in this new covenant must, as did Jesus, reclaim the Adamic vocational mandate to exercise benevolent co-regency with God in the world. The remainder of ch. 2 spells out some ramifications of this claim.

What Has the Second Adam Accomplished? (2:10-18)

With v. 10, the author of Hebrews points out some of the soteriological ramifications of the incarnation of the eternal Son as a human being. It is noteworthy that the section begins with a reaffirmation of God as creator, the one for whom (δι' ὅν) and through whom (δι' οὗ) all things (τὰ πάντα)

---

[15] N. T. Wright, *Simply Christian: Why Christianity Makes Sense* (New York: HarperCollins, 2006), ch. 1. "Putting the World to Rights" is the title of the first chapter of this volume, yet the language is nearly ubiquitous in Wright's writings and speaking engagements.

exist. This language is similar to and recalls the statements of the opening paragraph of the letter, where it was stated that through the Son (δι' οὗ) all things (τὰ πάντα) were made and that God appointed him heir of all things (1:2). This interchangeability of the attribution of creation to God and to the Son is certainly part of the fodder that fueled early Christian christological and Trinitarian reflection. But what of the assertion, in which God's act of creation is embedded, that it was "fitting" that God should perfect the pioneer of human beings' salvation through suffering?

The suffering of the incarnate Son seems somehow intrinsic to the logic of salvation, especially given that God is the one who is said to have created all things. Two points are of ecological import here. First, it appears that the scope of redemption includes far more than just human beings. True, the author of Hebrews restricts focus to the human component of the "all things," but given the statement that God is the creator of all things, it seems that the suffering of the Son is fitting in light of this. The author could have very easily omitted the reference to God creating all things and the sense would have been quite anthropocentrically framed. Yet the mention of God as creator of all things opens the prospect that the whole Earth community is properly in view with respect to the suffering of the eternal Son become human being. Whatever the Son achieves through suffering extends to all things.

Second, the fittingness of the Son's suffering points to the necessity of God's participation in creation. As we will discuss later, the *telos* of creation is to become the dwelling place of God. If God is to redeem creation, it would be appropriate to do so in a participatory way, entering into the brokenness and suffering of all of creation, doing so in such a way as to rehabilitate those who were to be the benevolent co-regents with God over God's good creation. Moreover, given that the Son's becoming human also connects him to the stuff of creation out of which the first human being was formed (cf. Gen. 2:7), the Son participates not only in human suffering but also in the suffering of the other-than-human creation that is the result of human failure to attain to its vocational destiny. The one who embodies the radiance of God's glory and is the exact representation of God's very being is fit for his redemptive mission of all things through suffering.

Hebrews 2:14 affirms that it is by becoming human that the Son identifies with the plight of human brokenness and the brutal reality that all of creation is subjected to the tyranny of death in its current state. As we noted earlier, human failure to exercise the vocational mandate of the original Adam has thrust all of creation into futility, into failure to realize its own destiny. Yet in the grand cosmic irony, it is by death that Jesus destroys the one who exercises the power and fear of death over human beings, namely, the devil. Once broken, the tyranny of death gives way to an approach to existence characterized by life. Here, again, the author focuses on how it is only by becoming like they are, flesh and blood, that the Son could share in the full human experience terrorized by the specter of death and thus procure atonement for the sins of the people (v. 17). But sharing in human physicality also identifies Jesus with the whole of creation and would entail some kind of benefit for the other-than-human order, if only indirectly via the reclamation of the Adamic vocation for now-forgiven human beings.

The author of Hebrews makes mention of two other consequences of the incarnation of the Son. First, the identification of the Son with human beings is the basis of a new familial relationship. Human beings are now the siblings (ἀδελφούς) with the Son for they have the same Father, God (Heb. 2:11), a status confirmed via three brief scriptural quotations (vv. 12-13). From an ecological perspective, this creation of a family through the redemptive work of the Son recalls the Adamic mandate to fill the world with offspring after the image of Adam, who is the image of God (cf. Gen. 1:28). This blessing of fruitfulness is connected in Gen. 1:28 with the exercise of dominion over all living things. Taken in conjunction with the quotation in Heb. 2:6-8 from Psalm 8, which is a reflection on Gen. 1:26-28, the formation of the new family with the Son, who is himself the exact representation of God's very being, entails reasserting the Adamic vocation, and by implication, the image of God and its ramifications in human beings.

Second, the high priesthood of the Son is mentioned for the first time. This will occupy our author's, and our, attention later. For now, we will simply foreshadow, much like the author of Hebrews does, the high priestly ministry of the Son. For the author of Hebrews, it is incumbent for the Son to share in our humanity so that he may be a merciful and faithful high priest and thus secure atonement for sins. For our ecological reading, the priesthood of the Son models the priesthood that was to characterize the vocation of human beings on behalf of Earth. This is derivative from his identification with us in our humanity. Ours was a priesthood designed to mediate God's benevolence to the rest of creation. In becoming human, the Son enters into the world to reestablish this vocation, a mission made possible through his priestly ministry on our behalf.

Here the concepts of family and priesthood converge. Under the old covenant, priesthood was established through familial lineage—a concept that comes under critique from the author of Hebrews. But perhaps here we may see a subversion of this particular critique of our author. The new family is in a sense the offspring effected through the Son's redemptive work. This is the restored human family, with God as its Father and the Son as its sibling. Father and Son are each connected in their own ways with the creation, sustenance, and destiny of all things. The lineage of Adam was to be the priesthood that would co-reign with God over creation. In the Son, the priestly lineage of human beings is restored in a way that has the quotation of Ps. 8:5-7 firmly in view as its goal. Again, these points will receive elaboration later. For now, we rest where the author of Hebrews has, foreshadowing what is to come.

Hearing Earth

In this section of Hebrews, Earth becomes more of a participant in the drama by virtue of its connection with human beings. Genesis 2:7 is a powerful tradition that cements an identification between the human and other-than-human components of the created order. Here Earth functions in a hortatory manner, urging human beings to recognize this identification and to let it govern how human beings understand their vocation within creation. From the beginning, human and other-than-human destiny have been intricately interconnected. Adam as exemplar of the benevolent priestly co-regent with God over all God created in the beginning, interrupted though it was for the expanse of human history due to human rejection of that vocation, has been restored in the Son who became human as we are in order to bear the suffering and break the curse of death over all things. The vocation has been reclaimed in the Son who is the exemplar of the new humanity and who is established as the high priest over a family of priests who exercise God's benevolent rule in the world. Earth proclaims loudly, "Take this mantle and own once again what is truly yours!"

This identification between human and other-than-human gives Earth the boldness to call out human activities that bring harm to the whole Earth community, including human beings. Given our shared substance, does it make sense to poison the ground with chemicals in order to grow food that will in turn poison human beings? How long will the descendants of Adam claim the right to exploit Earth without due consideration of the vocational mandate given to the first Adam? Can those who claim allegiance to the Second Adam, and who may even proclaim loudly in theological discourse to hold to the true humanity of the second person of the Holy Trinity, continue to deny the materiality of the incarnate Son by denying the material connection of human and Earth he embodies? Does Psalm 8 grant the children of Adam the right of convenience if it means polluting Earth's oceans with discarded plastics? Earth issues a wake-up call to the image-bearers not only to see the original vocational mandate but also to pledge fealty to the vocational mandate reclaimed in the incarnate Son. Human power and technology has come to a point where human beings can change Earth's ecology on a global scale. But is this a faithful way to exert the dominion granted human beings by their creator? Clearly they have retained the impetus to exercise dominion; the only question is whether they will do so in light of their vocational mandate or do so with concern only as to how this will serve their thriving in the world. Earth speaks as the conscience of human beings to remind them of where they began and where they have now come in the Son's redemptive work on their behalf.

4

Building a House on Earth (Heb. 3:1-19)

Contents

The previous section established the rationale for the Son becoming human. In sum, it was to restore human beings to the vocational mandate of the first Adam. In this section, that idea is extended. The Son, established by the author as superior to Moses, the exemplary figure of the old covenant, is said to be the builder of a house, which is nothing other than the people he became like in order to redeem. In reality, this functions to reclaim the mandate given to Adam to be fruitful and multiply and to fill the world (Gen. 1:28). But before commissioning this people for their task, they are warned not to do as the previous generation did. For the author of Hebrews, this is a call not to follow the wilderness generation into disobedience. From an ecological perspective, it may be seen as a warning not to forsake the Adamic vocation to serve as the priestly co-regents with God in creation as did the first humans.

Structure

3:1-6: What has the builder built?
3:7-19: The urgency of faithful response

Analysis

What Has the Builder Built? (3:1-6)

In Hebrews 3, the author turns attention to a further comparison of covenants by contrasting the Son, Jesus, with the figure most associated with the old covenant, Moses. Strictly speaking, this is not the first comparison with Moses, though it is the first time Moses's name is mentioned explicitly. The references in both chs 1 and 2 to the revelation of the old covenant clearly have Moses in view, for he is the prophet par excellence in Israelite history (cf. 1:1) through whom the angels mediated the message (cf. 2:2). Yet here the comparison is explicit, though in a way that moves the discussion beyond a comparison of revelations.

The comparison here is between Son and servant. Jesus is worthy of more honor due to his status as Son (3:6), whereas Moses, faithful though he was, attains to nothing more than a servant in God's house (v. 5). But there is a deeper issue here than just the relative rank of covenantal exemplars. The issue here revolves around the "house." In v. 2, Moses is said to be "faithful in all God's house." But crucial in the present argument is that more worthy of honor than a faithful servant is the builder of the house (v. 4). The author of Hebrews reiterates what was said in 2:10, that all things exist through and for God. Yet as we saw in 1:2, it was through the Son that God created all things. So the comparison between Moses and the Son runs deeper than mere status, servant versus Son. Rather, the issue is rooted in the doctrine of creation. Since God is the builder of the "house," and God did so through the Son, the Son ranks higher than Moses based on his role in the creation of all things.

Now the reader may be excused if the image of the tabernacle were to come to mind, for that is the house most identified with Moses. However, Moses is said to be one who portends that which is to come via the Son (3:5). In this regard, what does Moses portend? Verse 6 spells this out in two ways. First, the Son's authority over the house supersedes Moses's servanthood. As we saw in 1:3, the Son, once completing his redemptive mission, is seated in authority at God's right hand, exercising his rule, as we argued, through his people in the world. This hints that perhaps the house to which the author alludes is the "all things" over which the Son exercises authority. This notion that the house refers to the whole of the created order will find further support when we examine ch. 4.

Second, the author explicitly states that the house is the actual people of God (3:6). Christ's faithfulness is demonstrated in his rule within the house that is the people of God. This recalls the formation of the family from the preceding chapter. As we saw in that discussion, an aspect of the Son's redemptive mission was to reclaim for the posterity of Adam the role of benevolent priestly co-regency with God in creation, a role that casts the faithful as siblings with the Son and thus partakers with the Son, the Second Adam, of proper dominion over all things.

In an ecological framework, the connection of the house associated with Moses and the house(s) associated with the Son is suggestive. The notion of a "house" of God seems to indicate that this represents the place where God lives. As we will argue, Hebrews 4 boldly presents the idea that the whole of creation is designed to be the dwelling place of God. And yet, in 3:6, it is the people who are the dwelling place of God. Both ideas draw extensively on the tabernacle or temple imagery of the old covenant. And both contribute significantly to an ecological reading of the epistle. On the one hand, if the people of the new covenant are the dwelling place of God, then the original destiny of creation begins its realization, where God and human beings dwell together in God's creation. On the other hand, if creation itself is the dwelling place of God, then that establishes all of creation, human and other-than-human alike, as the objects of God's redemption in the Son. Both ideas are necessary for a fully orbed understanding of the place of Earth in God's economy.

At this stage, these ideas are anticipatory. A major step in this understanding will occur in ch. 4. But first, another warning occurs. The following warning allows for a framing in ecological terms that further reinforces the vocational mandate of human beings while it stands as a prophetic call for human beings not to forsake this mandate.

The Urgency of Faithful Response (3:7-19)

In my earlier volume, I characterized the Holy Spirit as a "whispered voice in the choir."[1] The Holy Spirit is mentioned explicitly in Hebrews in five instances (2:4; 3:7; 6:4; 9:8; and 10:15), with two other likely references (9:14; 10:29). Of interest in that discussion was to tease out a thematic ecological pneumatology in Hebrews, and the discussion developed as more of a theological rather than biblical assessment of the ecological import of the Holy Spirit in Hebrews. Our interests here, as pertains to the Holy Spirit, will remain quite narrowly concerned with the present passage and the role of the Spirit in the warning.

---

[1] Lamp, *The Greening of Hebrews?*, ch. 5.

The preceding paragraph concluded in 3:6 with a caveat that we are indeed the house of God if we hold fast to the confidence of our hope. With v. 7, the author draws a conclusion (Διό, "therefore"), attributing the following quotation from Ps. 95 [94 LXX]:7b-11 to the direct communication from the Holy Spirit to the author's present audience. The present urgency of the Spirit's message to the audience is highlighted by two features in Heb. 3:7.[2] The first is the present tense verb λέγει ("says"). This draws attention to the present activity of the Holy Spirit in the community. The fact that the Spirit currently speaks through the words of a psalm that is several centuries old by this time gives the warning in the psalm immediacy. Second, the author of Hebrews repeatedly makes use of the word "today" (σήμερον, v. 7) in the rest of the chapter (vv. 13, 15) to highlight the urgency of the warning to the present circumstances of the audience. This will prove important as the letter moves into ch. 4, but here it adds to the urgency of the warning.

The author's use of this psalm draws from the LXX, which serves the author's purpose well. It is interesting that in the Hebrew text of Ps. 95:8, the psalmist makes reference to two place names where the Israelites complained to Moses about the lack of water, Massah and Meribah, referring to incidents recorded in Exod. 17:1-7 and Num. 20:2-13, respectively. The reference to the people not entering God's rest in Ps. 95:11 refers to God's pronouncement at Kadesh (Num. 14:20-23), where God declares that the people's refusal to enter the promised land ensures that none of that generation will enter into the promised "rest." For the author of Hebrews this, of course, is a prelude to the discussion that will occur in Hebrews 4. At this point, however, it is worth noting that in the LXX, the translator chose to render the place names as ὁ παραπικρασμός ("the rebellion") and ὁ πειρασμός ("the testing"), respectively. William Lane suggests that the LXX translator is interpreting the psalm in light of Numbers 14, so that the rebellion and day of testing refer to the people's refusal to enter the promised land.[3] The author of Hebrews draws on the LXX translation, then, to make application of the psalm more pertinent to the circumstances of the present audience. Mention of Massah and Meribah are not pertinent to the situation facing the audience.

---

[2] Lamp, *Greening of Hebrews?*, 53.
[3] William L. Lane, *Hebrews 1-8*, WBC (Dallas: Word, 1991), 85-86.

So what is the substance of the warning? The author warns the readers not to follow in the way of the wilderness generation. Interestingly, this is the warning the psalmist originally provided his own readers centuries earlier: do not follow in the path of that generation that preceded them. The warning is timeless, always aimed at the covenant people, exhorting them always to be faithful and obedient. Mere status as the covenant people is no guarantee that they will receive the promises of the covenant.

How might we read this warning ecologically? Clearly, it is a warning aimed at the response of the people to God's activity among them. Be faithful, be obedient, or else face the consequences of unbelief and disobedience. A strong anthropocentric element is observable. Yet the author's choice of the LXX version of the psalm may open doors for an ecological reading. As noted earlier, the choice of the LXX version makes the warning more general in its application. It is this general tenor in the quoted psalm that allows for a subversive reading that forges a point of identification with Earth. In vv. 16-18, the author asks a series of questions designed to make the audience aware of their own susceptibility to failure. But in v. 17, the detail is given that God was angry with them for forty years. It was not just the incidents at Massah, Meribah, and Kadesh over which God was angry; it was for the forty-year span after God delivered the people from Egypt. So what happened during this period that may have ecological significance?

Almost immediately following the people's redemption from Egypt, the episode of the forging of the golden calf occurs (Exodus 32). As we noted in the previous chapter of our study, the true nature of the sin in Genesis 3 was human beings' abdication of their role as the priestly co-regents with God over creation, submitting themselves to the command of the other-than-human order over which human beings were to exercise benevolent dominion. Idolatry is, in one sense, simply nothing more than human beings replacing God with creation. The book of Wisdom (15:14-19) provides a scathing denunciation of the depravity of Egypt's idolatry, which is their submission to deities that are drawn from the animal kingdom.[4] Paul, in Rom. 1:18-25, makes similar statements against the futility of idolatrous worship of animal images. This

---

[4] Lamp, *Reading Green*, 101–6. See also Jeffrey S. Lamp, "Wisdom Pneumatology and the Creative Spirit: The Book of Wisdom and the Trinitarian Act of Creation," *Spiritus: ORU Journal of Theology* 2:1–2 (2017): 39–56.

type of idolatry subverts God's order and humanity's place within it. When the Israelites crafted the golden calf in the wilderness, they adopted for themselves the very practices for which Egypt was judged—the practices that ultimately led to their demise and Israel's liberation.

The warning here, then, is able to draw attention to an episode that contributed to the LXX translator's interpretation of Psalm 95 in order to include a heinous act of idolatry that is indicative of the abdication of humanity's ordained role within God's orders of creation. Moreover, the author of Hebrews invited this observation by extending the scope of God's displeasure with the Israelites to the entirety of their wilderness wanderings. Read ecologically, the warning of Heb. 3:7-19 is a vital one that addresses the immediate urgency of the psalm's warning to remain obedient to God. Here, we may insert one more area in which human beings are to remain faithful and obedient: the performance of their role as benevolent priestly co-regents with God in creation.

Hearing Earth

Hearing Earth in this section requires adopting the sense of urgency with which the author of Hebrews is concerned. So dire is the immediate circumstance that it is the Holy Spirit who speaks on behalf of Earth. We will speak more of the Spirit when we discuss Heb. 9:14, but here we focus on the Spirit giving voice on behalf of Earth.

The Spirit speaks first in the noted sense of urgency on behalf of Earth. "Today," says the Holy Spirit, if you hear God's voice at all in the words of the quoted psalm, do not become as those who rebelled and resisted God in the wilderness. Specifically, do not follow in the way of the first Adam and forfeit your role to become once again benevolent priestly co-regents with God in creation. Do not fail to assume the priesthood in creation that is now available in the priesthood of the Son. Do not fail to become the house in which God lives, through which God will come to inhabit Earth as the eschatological dwelling place of God. This must happen "today." As with all areas in which human beings are to be faithful and obedient to God, procrastination is not a viable option.

The Spirit, speaking on behalf of Earth, groans along with Earth in its suffering (see Rom. 8:18-27). Can the spiritual progeny of the Second Adam neglect its reclaimed vocation to serve and protect Earth? How long can Earth's oceans absorb increasing concentrations of carbon dioxide until acidification threatens aquatic life beyond its ability to thrive or adapt? Or before that, will the human consumption of single-use plastics reduce the oceans to watery garbage dumps, literally choking out organisms that mistake the waste for food? Will human agricultural practices strip Earth of its fecundity? Will climate change threaten life in the Earth community, human and other-than-human alike, on a global scale? How long can the human co-regents trade their rightful dominion for crass exploitation before Earth's ecosystems cross thresholds of no return? Earth, through the Spirit, cries out, "Today!" as it calls on human beings to respond. For if response is not quickly forthcoming, Earth will be forced to join God in judgment on human beings for their idolatrous exploitation of that over which they were to rule benevolently with God.[5] There is no time for business as usual. The time is now to act, today!

---

[5] See Lamp, *Reading Green*, ch. 5, for discussion of this notion.

5

Establishing Rest (Heb. 4:1-13)

Contents

As we have observed, this ecological reading of Hebrews so far has largely been a mapping of the work of the Son in creation and redemption onto the creation narratives of Genesis 1 and 2. The Son, involved in the work of creation, comes to restore the descendants of Adam to the vocation of Adam by becoming the Second Adam who fulfills that vocation and passes it on to the covenantal descendants of this Second Adam. In this section, we exercise a vivid subversion of the author's apparent intention. The focus here seems to be on encouraging the first readers to remain faithful and thus enter into the experience of salvation, characterized here as the Sabbath rest of God (cf. Gen. 2:2-3). In an ecological reading, the focus shifts from a more passive reception of salvation to the active effecting of this salvation for Earth. The focus is not the mere comparison of "Jesuses," but the commissioning of the people created to reclaim their Adamic vocation, as described in the previous section.

Structure

4:1-11: A people of God's Sabbath rest
4:12-13: Another warning to accountability

Analysis

A People of God's Sabbath Rest (4:1-11)

The warning delivered, the author now proceeds to exhort the people to faithful obedience so that they may not follow the example of the wilderness generation and find themselves excluded from the promises of the covenant. The imagery employed by the author is Joshua's entrance into and conquest of the promised land. Coming as it does on the heels of ch. 3's comparison of the Son with Moses and the rebellion of those who wandered in the wilderness, reference to Joshua is appropriate. The idea here is that the "rest" that is available in the Son is superior to the rest that was available through Joshua. The advancement in the argument here is that the rest remains open to those of the author's day to be entered through faithful obedience. This conclusion is reached on two bases. First, in Heb. 4:7, appeal is made to David in the quotation from Ps. 95:11. The psalm was written "much later" (μετὰ τοσοῦτον χρόνον) than the time of the rebellion in the wilderness, which to the author of Hebrews indicates that the promise to enter God's rest was open after Joshua's time. Second, the reason that the promise of rest remains available in the author's day is due to the failure of the rest under Joshua to be fully actualized (Heb. 4:8). If Joshua did not give the people rest, and David affirms that the promise of rest remained open later than Joshua's day, then it is reasonable to assume, in the author's estimation, that the promise of rest remains open "today" and indeed finds its realization in the Son.

In my earlier volume, I noted that a strong anthropocentric bias is discernible in the argument of Heb. 4:1-11.[1] On the one hand, the land is not treated as a subject of integrity in the discussion. The promise of the land, a significant aspect of the promise given to Abraham, is reinterpreted by the author of Hebrews such that it is in the Son that it finds its realization. In the Hebrew Scriptures, the land is the subject of a great deal of Pentateuchal legislation, most notably in Leviticus 25–27. In his study of the land in ancient Israel, Norman Habel identifies the focus of Leviticus 25–27 as reflecting an agrarian ideology in which the land is depicted as God's garden and the people of Israel as tenant farmers of God's garden.[2] Of particular note is that the land itself is the recipient of Sabbath rest, as the following passage from Exod. 23:10-12 (cf. Lev. 25:1-7) exemplifies:

> For six years you shall sow your land and gather in its yield; but the seventh year you shall let it rest and lie fallow, so that the poor of your people may eat; and what they leave the wild animals may eat. You shall do the same

---

[1] Lamp, *The Greening of Hebrews?*, 38–42.
[2] Norman C. Habel, *The Land Is Mine: Six Biblical Images* (Minneapolis: Fortress, 1995), ch. 6.

with your vineyard, and with your olive orchard. Six days you shall do your work, but on the seventh day you shall rest, so that your ox and your donkey may have relief, and your homeborn slave and the resident alien may be refreshed.

The Sabbath year rest for the land follows on analogy for the weekly Sabbath day rest experienced by the people. In addition, every fifty years, the land would gain an extra year of rest in the Year of Jubilee (Lev. 25:8-12). As the remainder of Leviticus 25 indicates, this legislation regarding the treatment of the land is an integral component of the webs of relationships that exist among human beings and the world around them. Leviticus 26:3-45 spells out the ramifications of the people's observance of all that God has commanded them, including their observance of Sabbath legislation. Should they follow God's commands, then the land will be fruitful and produce what they need to account for the time in which the land was unproductive. However, should the people prove disobedient and fail to give the land its rest, God will expel the people from the land so that the land will enjoy the Sabbath rest that was denied it (Lev. 26:34-35).[3]

What we see in Heb. 4:1-11, however, is a reinterpretation of Sabbath rest that turns the focus away from the land and focuses it on a spiritual fulfillment of the promise in the Son (v. 3). The promise of the land that was ostensibly available through Joshua is no longer a promise of land, but a christologically centered reality that speaks only to the thriving of human beings. The notion of rest (κατάπαυσις) is brought forward from the citation of Psalm 95 in ch. 3 and sets the stage for the discussion in Heb. 4:1-11. However, in v. 3, the focus has turned from the rest Joshua brought to God's work in creation. The author notes that those who are currently faithful indeed enter into that rest. As I noted in my earlier volume, the author here engages in a piece of Jewish exegesis known as *gezerah shawa*, an interpretive technique that joins together passages based on common vocabulary, to define the rest not in terms of

---

[3] Fred Van Dyke has argued that the prophecy of Jeremiah that the people would remain in exile in Babylon for seventy years refers to the Israelites' failure to allow the land its Sabbath years. He cites 2 Chron. 36:20-21: "[Nebuchadnezzar] took into exile in Babylon those who had escaped from the sword, [. . .] to fulfill the word of the Lord by the mouth of Jeremiah, until the land had made up for its sabbaths. All the days that it lay desolate it kept sabbath, to fulfill seventy years." See Fred Van Dyke, *Between Heaven and Earth: Christian Perspectives on Environmental Protection* (Santa Barbara: Praeger, 2010), 62.

Joshua's conquest but in terms of God's work in creation.[4] In this case, the noun from Ps. 94:11 (LXX), κατάπαυσις, is connected with the cognate verb κατέπαυσεν ("rested") in Gen. 2:2 (LXX). In the Genesis passage, God is said to have rested from God's works, having assessed them "very good" (Gen. 1:31). The effect in Hebrews is that the rest that is available "today" in the Son is what Fred B. Craddock termed "participation in God."[5] The rest to which the psalmist alluded was not truly the land; it was a quality of life available only in the Son who is the embodiment of God's very nature. The focus is no longer on responsible living in the land of promise; it is on how human beings experience the blessings of divine life in the Son for which Joshua served merely as a shadow. Sabbath is no longer a rubric for the web of interactions between God, human beings, and the other-than-human order; it is a rubric for the reward available to those who remain faithful and obedient.

An ecological reading of the passage is possible through a subversion of the author's reinterpretation of the promise of the land. Crucial here is a recovery of the meaning of Sabbath rest, especially as it is brought into connection with God's work in creation. Here the author of Hebrews has provided an invaluable service for us by connecting the notion of Sabbath rest with the doctrine of creation (4:3-4). The collocation of these concepts provides the foundation for an ecological reading of Heb. 4:1-11.

The first of the two creation accounts in the opening chapters of Genesis, 1:1–2:3, concludes with the assertion that on the seventh day, God rested from the work of creation (Gen. 2:2). In Heb. 4:10, the author characterizes the Sabbath rest awaiting the people of God as a cessation of labor modeled on God's rest from the works of creation.[6] Such an understanding apparently

---

[4] Lamp, *Greening of Hebrews?*, 42–43.
[5] Fred B. Craddock, *Hebrews*, NIB (Nashville: Abingdon, 1998), 52.
[6] There is some debate on whether the blessing of Sabbath rest is cast as an eschatological or a present reality. For those who see an eschatological reality, appeal is made to the characterization of the present life as a pilgrimage toward this goal (Heb. 4:11) and the Sabbath rest as a cessation from labor (4:10). Examples of those holding this view are Luke Timothy Johnson, *Hebrews*, NTL (Louisville: Westminster John Knox, 2006), 130–31; Craig R. Koester, *Hebrews*, AB (New York: Doubleday, 2001), 270, 279; Paul Ellingworth, *The Epistle to the Hebrews*, NIGTC (Grand Rapids: Eerdmans, 1991), 246; Harold W. Attridge, *The Epistle to the Hebrews*, Hermeneia (Philadelphia: Fortress, 1989), 131. Those who see it as a present reality appeal to the present tense verbs in vv. 3, 6, and 9, and the use of "today" in v. 7 and in the preceding context (3:7, 13, and 15). Examples of those holding this view are Ben Witherington, III, *Letters and Homilies for Jewish Christians* (Downers Grove: IVP Academic, 2007), 177, 182; Lane, *Hebrews 1–8*, 99; Craddock, *Hebrews*, 52; Robert Jewett, *Letter to Pilgrims: A Commentary on the Epistle to the Hebrews* (New York: Pilgrim Press, 1981), 66–67; Schenck, *Cosmology and Eschatology in Hebrews*, 60–63. Regardless of the decision

gave rise to the coinage of a new term for this rest, σαββατισμός—a term that differs from κατάπαυσις by virtue of its connection to God's rest at the conclusion of the work of creation. The author's exegesis here gives the proof of the superiority of the Son's bestowal of rest over that of Joshua. The true nature of rest is understood only in connection with divine life, life only available through the Son who is the embodiment of God. The fact that in Greek Joshua's name is identical to that of Jesus (Ἰησοῦς) provides literary poignancy to the author's argument. The first Jesus was only always a precursor, a shadow, of the coming Jesus who would grant the gift of σαββατισμός, a rest far superior to the mere κατάπαυσις available through Joshua.

It is precisely at this point our subversive reading enters the picture. We need to take the bait given by our author and probe more deeply into the Genesis tradition employed in Heb. 4:1-11. In more recent scholarship, several studies have made the comparison between this first Genesis creation account and the accounts of the construction of the tabernacle or temple. William P. Brown, for example, has compared the structure of Gen. 1:1–2:3 with Exodus 25–40, where the construction of the tabernacle is described in great detail.[7] Brown concludes that the structural parallels between the two accounts are not mere coincidence, but rather indicate that the two accounts are to be understood together. The idea here is that the purpose and function of the tabernacle, and later the temple, is to serve as part of God's project of redemption to restore all of creation to its intended purpose, namely, to be the dwelling place of God.[8] If this is indeed the case, then what Gen. 2:1-3 contributes to the first creation account is the coming of the deity to inhabit the completed temple.[9]

Margaret Barker has applied her understanding of biblical theology as "temple theology" to the current environmental crisis in her monograph,

---

here, the point is that the focus has turned from seeing the land as a subject in its own right to seeing a spiritual reinterpretation of the promise in the Son.

[7] Brown, *The Seven Pillars of Creation*, 47. Margaret Barker provides a description of correspondences between the creation account of Genesis 1 and the vision Moses had on Sinai regarding the construction of the tabernacle in *Creation: A Biblical Vision for the Environment* (London and New York: T & T Clark, 2010), 38–49.

[8] See Gregory K. Beale, *The Temple and the Church's Mission: A Biblical Theology of the Dwelling Place of God* (Downers Grove: IVP, 2004), 29–167; Jonathan Huddleston, *Eschatology in Genesis*, Forschungen zum Alten Testament 2. Reihe (Tübingen: Mohr Siebeck, 2012), 157; Laansma, "Hidden Stories in Hebrews: Cosmology and Theology," 9–18; John H. Walton, *Ancient Near Eastern Thought and the Old Testament: Introducing the Conceptual World of the Hebrew Bible* (Grand Rapids: Baker, 2006), 113–34.

[9] N. T. Wright, *Paul and the Faithfulness of God* (Minneapolis: Fortress, 2013), 102, 560.

Creation: A Biblical Vision for the Environment. Germane to her argument is the contention that current readers of the Bible "return to the sources in the quest for biblical creation theology, which for Christians must be the Bible as it was and as it was understood in the time of Jesus."[10] Her exposition attempts to reconstruct what Christians "could have known" about creation in order to frame a response to current environmental crises in terms of creation as the temple of God.[11] Her study is a thorough examination of biblical and extrabiblical creation texts, resulting in a picture of creation as God's temple and Adam as the priest in this temple. While her thesis has been critiqued on methodological grounds for its attempt to address present ecological crises through a coherent biblical theology that casts creation in terms of temple imagery,[12] connections are possible and suggestive in our ecological reading of Hebrews.

More pertinent to our discussion here than scholarly investigations into the historical roots of the identification of creation with the tabernacle or temple is the strategic move of the author of Hebrews in bringing together the notion of Sabbath rest with creation. Clearly missing from our author's discussion is the Sabbath legislation regarding the land. Rather, appeal is made to the creation account in Gen. 1:1–2:3. Yet it is clear that Pentateuchal Sabbath legislation regarding the land is derived on analogy with the affirmation that God rested on the seventh day. While the application of the Genesis account in Heb. 4:1-11 is a spiritualized one, the tactic by which the author of Hebrews achieves this

---

[10] Barker, *Creation*, 13.
[11] Barker, *Creation*, 32–33.
[12] John Painter has subjected the volume to serious critique, concluding that Barker has made too facile a leap from what biblical writers *could* have known about creation to what they *did* know, leaving Painter unconvinced about the biblical grounding of Barker's vision. Painter's critique raises significant issues regarding Barker's methodology. This may seem to deal a decisive blow to our reading of Hebrews because our ecological reading of Hebrews to this point largely arrives at the same conclusions as Barker. However, our methodological approach differs. Rather than try to establish on historical and exegetical grounds that Jesus and the biblical writers had worked out a coherent creation theology that directly impinges on today's environmental crises, we acknowledge that such would be difficult to demonstrate precisely because the exigence of anthropogenic environmental catastrophe on the scale of today's crises simply did not exist at the time. What we are arguing is that, given the starting point of today's ecological crises, we may approach biblical texts from an ecological perspective and ask the Bible questions that would not occur to the writers and earlier generations of readers, teasing out implications for our concerns. More succinctly put, we openly acknowledge that Earth's concerns are not primary for biblical authors, but our concerns may provide opportunity for teasing out implications that may not have occurred to those living at the times of the composition of these texts. See John Painter, "Review of Margaret Barker, *Creation: A Biblical Vision for the Environment*," *Review of Biblical Literature*, 2011. Available online: http://www.bookreviews.org (accessed July 25, 2019).

application plays into our hands. Sabbath rest is indeed an aspect of the created order. The first creation account is incomplete by the end of Genesis 1. There is no indication of the destiny of creation. Why is it here? The answer seems clear: God wishes to indwell what God created in order to share in life together with this creation. Numerous indications throughout the biblical drama attest to this, beginning in Genesis itself. In Gen. 3:8 it is said that God walked in the garden in the evening breeze. Book-ending this account is the vision in Rev. 21:22 where God indwells the created order, alleviating the need for a temple due to God's very presence among the people in the newly descended heavenly Jerusalem.

The author of Hebrews has virtually invited reconsideration of the creation account with the discussion of Sabbath rest grounded in the Genesis creation account. Of special interest here is the unvoiced question that the author's own exegesis raises. Granted that there will be a Sabbath rest awaiting those who remain faithful and obedient to the Son, where will they experience this rest? We will have more to say about this when we survey Hebrews 11 and 12, but some clue has already been given. In 1:2, the Son is said to be heir of all things. Everything the author has asserted about the Sabbath rest available in the Son may indeed be just as argued, but there is still more to say here. The people still need to be somewhere. Of course, the option exists that perhaps their final abode may be heaven following the destruction of the physical cosmos. We have presented some preliminary objections to this prospect, and we will offer more later. Joshua's rest involved a place for the people to be. Perhaps the Son's rest, too, involves a place to be. Joshua's rest involved a particular piece of geography. The Son's rest may far exceed this to include the whole of the created order. This is more likely if our reflections on the purpose of the original creation extend to address its ultimate destiny.

Another Warning to Accountability (4:12-13)

This brief warning functionally closes out this section of the epistle. In what is one of the most well-known verses in the epistle, v. 12, the author speaks of the piercing power of the "word of God" to judge the deepest thoughts within human beings. In its context, the verse serves as a clear warning to remain faithful, because the word of God exposes all secrets and holds all accountable. It is tempting to identify precisely the referent for the "word of God." Given the sheer quantity of Scripture cited in the letter so far, it may be that the Scriptures in general are in the author's mind here. Or more immediately, it may be that the passage cited from Psalm 95 is in view. Moreover, interpreters seek for clues to the author's anthropology in the mention that the word of God functions like a surgical instrument able to separate even soul (ψυχή) and spirit (πνεῦμα). I have argued elsewhere that perhaps more significant than these items of speculation is the qualifiers "living" (ζῶν) and "active" (ἐνεργής).[13] With the mention of the Holy Spirit speaking to the audience (3:7) and the repeated refrain "today," the author seems more concerned with the urgency of the warning than the precise details within the warning.

From an ecological perspective, however, we may benefit by focusing on the scriptural allusion to Gen. 2:2. If our reflections on this passage have merit, the urgency of the warning here meets with the call to consider more fully the implications of what is meant by Sabbath rest. Yes, it is important not to follow the footsteps of the wilderness generation and thus miss out on this rest, but it is also important to consider the fuller import of what may be entailed by the concept of Sabbath rest, especially as it is connected with God's work in creation. If the destiny of creation is to be the dwelling place of God with human beings functioning in the priestly co-regency recapitulated in the Son, the Second Adam, then how we respond to this reality becomes subject to the penetrating examination of the Holy Spirit as we are confronted with the word of God in this regard today. In truth, the scope of the warning is more expansive than the author of Hebrews probably envisioned. Remaining faithful and obedient now includes proper response to the newly regained vocational mandate to serve and protect Earth.

Hearing Earth

There are several areas in which Earth may speak to us from this section of the epistle. First, and most pertinent, Earth calls to us to reverse engineer the exegesis that led to the author's spiritualization of the promise of the land in the Son. As we saw, Sabbath is a concept that was wrested from its implications for the land to speak only on an anthropocentric level. This was achieved by the

---

[13] Jeffrey S. Lamp, *The Letter to the Hebrews: A Centre for Pentecostal Theology Bible Study* (Cleveland: CPT Press, 2017), 42.

author's linking of Ps. 94:11 (LXX) with Gen. 2:2 via the technique of *gezerah shawa*. As I suggested in the earlier volume, what might be achieved if we were to use this technique to make connections from the author's use of Sabbath in Heb. 4:1-11 and passages that use the term in the Hebrew Bible?[14] One such effect would be the reclamation of the Sabbath regulations regarding the land as an aspect of enjoying God's Sabbath rest in the Son. In fact, Earth might say that this exercise would only confirm what is already implicit in Jesus's own understanding of his mission in what has been called his "Nazareth Manifesto" in Lk. 4:16-21.[15] In the final line of the passage read by Jesus from Isa. 61:1-2 (Lk. 4:19), he makes mention of the "year of the Lord's favor," a reference that has in view the Year of Jubilee. The Jubilee legislation, which parallels the Sabbath legislation in Leviticus 25, contains provisions for the well-being of the land.[16] Earth makes the claim that re-extending Sabbath to the actual land is not inimical to the Sabbath rest available in the Son, but is integral to it.

Perhaps another way Earth might speak to us regards the Son's superiority to Joshua. Under Joshua the land was parceled and divided among people. In the Son, there is no division for it is in him that human beings find unity and their "landedness." Participation in the Son is the true home, the true Sabbath rest, in which land is not used as a commodity, resource, or source of division. Moreover, land itself is taken up into Sabbath rest and finds its own destiny in the Son. Divine/human/other-than-human communion is established in the one who is himself the capitulation of humanity, but is also the capitulation of the stuff of Earth from which humanity derives its own existence.

One last way Earth speaks through this passage is to draw attention to the destiny of creation—a consideration introduced in the author's appeal to Gen. 2:2. If indeed creation is destined to become the dwelling place of God, in accordance with creation's original design, human beings must reclaim their destiny to be God's priestly co-regents in and for Earth. Here the sense of urgency entailed in the word "today" and in the Spirit-inspired call through the word of God focuses attention on the critical issue of re-appropriating the promise of Sabbath for Earth. The Earth community does not have the luxury of time. According to projections by the IPCC, long term effects to

---

[14] Lamp, *Greening of Hebrews?*, 47–48.
[15] Nick Spencer, Robert White, and Virginia Vroblesky, *Christianity, Climate Change, and Sustainable Living* (Peabody: Hendrickson, 2009), 102.
[16] Lamp, *Reading Green*, 42–49.

Earth's ecosystems from anthropogenic greenhouse gas emissions are virtually certain, even if emissions of these gases are dramatically curtailed in the near future, and indeed effects are already being felt.[17] Earth may indeed be destined for eschatological transformation into the dwelling place of God, but eschatology must also serve as a driver for the proleptic anticipation of Earth's future in the present.[18] It can never be enough to wait passively for this transformation; it is always the design of Earth to be God's dwelling place as it is always the vocation of human beings to be God's priestly co-regents in creation. For Earth to attain to its destiny, it must be on the path to attaining it in the present. This demands an immediate response from human beings.

---

[17] See IPCC, *Climate Change: Synthesis Report. Contribution of Working Groups I, II and III to the Fifth Assessment Report of the Intergovernmental Panel on Climate Change*, eds. R. K. Pachauri and L. A. Meyer (Geneva, Switzerland: IPCC, 2014); IPCC, *Global Warming of 1.5°C. An IPCC Special Report on the Impacts of Global Warming of 1.5°C above Pre-industrial Levels and Related Global Greenhouse Gas Emission Pathways, in the Context of Strengthening the Global Response to the Threat of Climate Change, Sustainable Development, and Efforts to Eradicate Poverty*, eds. V. Masson-Delmotte et al. (Geneva, Switzerland: World Meteorological Organization, 2018).

[18] N. T. Wright calls this "eschatology in the process of realization" in *Surprised by Hope: Rethinking Heaven, the Resurrection, and the Mission of the Church* (New York: HarperOne, 2008), 208.

6

A New, Yet Ancient, Priesthood Introduced (Heb. 4:14–6:20)

Contents

The author proceeds to show how Jesus the Son is a superior priest to the Levitical priesthood established under the old covenant. Crucial to the argument is that Jesus is able to sympathize with us because he is human like us, yet he is so much more. The enigmatic figure of Melchizedek is brought into the discussion as the template for this new priesthood. The author of Hebrews introduces Melchizedek for purposes upon which ch. 7 will elaborate. Before delving more deeply into the nature of Jesus's priesthood, the author interjects a somewhat caustic criticism and warning to readers as well as an appeal to God's faithfulness as seen in the covenant with Abraham. Hebrews is replete with warnings to adhere to the new covenant, with each being more dire than its predecessor. Then the author introduces Abraham as a strategic player in the comparison of Jesus's priesthood to that of the old covenant. In this regard, the present section furthers the argument that the new covenant surpasses the old. Yet from an ecological perspective, the introduction of a new, yet ancient, priesthood serves another purpose. Having become a priest on our behalf, his identification with us forms the basis for conferring the reclaimed Adamic vocation to those who are in the new covenant. Jesus is the new Adam; his spiritual descendants will be imaged after him and function in his priestly vocation. The readers are then exhorted to orient themselves that they might ponder the significance of how Jesus's priesthood prepares them to enter more fully into the Adamic priestly vocation. Then, rather than functioning as a foil for the author's comparison of conveantal priesthoods, Abraham is introduced as a pivotal figure in the Son's ministry of reclaiming the Adamic vocation for human beings. This all prepares readers to consider the depths of Jesus' priesthood in the next chapter.

Structure

4:14–5:10: Human like us to be a priest for us
5:11–6:12: A charge to enter more deeply into this priesthood
6:13–20: Abraham and God's covenantal faithfulness

Analysis

Human like Us to Be a Priest for Us (4:14–5:10)

This section begins with a simple conclusion drawn from the preceding argumentation. Because "we have a great high priest who has passed through the heavens, Jesus, the Son of God, let us hold fast to our confession" (Heb. 4:14). Here we see three appellations assigned to the initiator of the new covenant. He is called a high priest, reintroducing the discussion temporarily abandoned from ch. 2 and with which the author will briefly entice us before interrupting us temporarily once again. Second, the name "Jesus" recalls the immediately prior discussion where he was favorably compared to the other "Jesus," Joshua. Finally, he is called the Son of God, recalling the focus of the opening chapter of the letter. But at this point, two features of this high priest are highlighted. First, he has passed through the heavens. This fact is the foundation for the exhortation to hold fast to our confession. This is indeed the one who was established as eternal in ch. 1, but here the Son's eternality is connected to his high priesthood. This move will prove crucial in ch. 7. Here it simply moves the discussion forward. Our high priest is at home among the heavens.

Another aspect of the Son's priesthood is brought forward here. In Heb. 4:15 the author asserts that the Son is suitably fit for his role as our high priest because he is able to sympathize with our weaknesses. In 2:10, the author argued that the Son had to be made perfect for his redemptive role through suffering as a human being and it was for this reason that he had to become human like us. Here, the argument becomes more poignant. The Son becomes like us in our weaknesses as frail human beings subjected to sin and death, with the qualifier that he did so without sin. This becomes a crucial issue for our ecological reading. It is not the case that the Son came to *attempt* to do what the first Adam could not. The Son, in facing the weaknesses of human frailty, came to recapitulate Adam's vocation. From earliest times, Christian theology has wrestled with the question of whether Jesus was truly capable of failing the test, as did Adam. Such speculation is not the author's interest here. Rather, the issue is that Jesus has faced our human predicament yet did not fail as did Adam. Jesus is the restoration of the race of Adam, replete with all that the first Adam's vocation entailed.

In the opening paragraph of ch. 5, the author turns attention to the nature of priesthood as depicted under the old covenant. Human high priests are selected from among human beings and are appointed to offer sacrifices for sins on behalf of human beings. But crucial to the performance of their duties is the fact that the high priests themselves are subject to weakness and are thus sinful. This grants them the capacity for sympathy with the people on behalf of whom they minister, for they must also make sacrifice for their own sins prior to offering sacrifice for the people. Here the author mentions another name associated with the old covenant: Aaron (v. 4). If Jesus is superior to Moses and Joshua, exemplars in their own rights of the old covenant, then he must also be shown superior to Aaron.

The author here is demonstrating not only that Jesus is superior to Aaron but also that Jesus's priesthood follows in the pattern of Aaron's priesthood, at least substantially so. Jesus is indeed able to sympathize with us because he, as human like us, is subject to the weaknesses that characterize human existence. So Jesus must, like those in the priestly line of Aaron, experience those very things, yet triumph over them. But also like Aaron, Jesus does not aspire to the status of high priest of his own will. Rather, he must be appointed, and here the unique nature of his high priesthood is demonstrated with appeal to two scriptural citations. First, Jesus is established as the Son of God. With a quotation from Ps. 2:7, which the author previously cited in Heb. 1:5, the author affirms that Jesus is appointed to his position of power by the declaration that he is indeed God's Son.[1] But the author here connects this ascription

---

[1] William L. Lane, for example, is correct to see this not as a designation of parentage but of power and authority. See Lane, *Hebrews 1–8*, 118.

of power with a citation from Ps. 110:4, where the Son is declared to be of another priestly order, that of Melchizedek (Heb. 5:6). Aaron was appointed to his priesthood by the will of God. Jesus is appointed to a qualitatively higher order of priesthood, also by the will of God.

In what is possibly an allusion to the Gethsemane event,[2] Heb. 5:7 highlights the Son's susceptibility to human frailty and thus his identification with human beings in their weaknesses. This, again, is identified as that which enables the Son to procure eternal salvation for human beings. But the author circles back around to connect this to the Son's priesthood in the order of Melchizedek. So how does this all inform an ecological reading of Hebrews?

To appropriate the vocabulary of ecological hermeneutics employed by Norman Habel and the Ecological Hermeneutics Section of the Society of Biblical Literature, the emphasis here is on identification, though in a slightly modified sense. In Habel's formulation, identification refers to an empathetic connection with Earth. Here, we see this in an indirect way. The author of Hebrews goes to great pains to show how the Son is qualified to serve as our high priest because he endured human weakness as have human beings, though without sin. As we will see in ch. 7, this enabled him to offer a once-for-all sacrifice that accomplished what the old covenantal sacrificial system was unable to achieve. But viewed from an ecological angle, the Son's identification with human weakness contributes to an ecological reading of Hebrews via the original Adamic vocation to serve as priestly co-regents with God over creation. The two scriptural citations in vv. 5-6 function to establish the Son as an eternal high priest in the order of Melchizedek. These citations firmly ground the Son's priestly authority in his identification with human weakness. But the difference here is that the Son fulfilled human destiny by not committing sin in his fleshly existence. In effect, the Son recapitulated the Adamic vocation by being appointed to a priesthood grounded in divine authority. So on one level, the Son forged an identification with Earth by reestablishing human beings to their proper status as those who serve as priestly mediators between God and Earth.

---

[2] Lane goes to great effort not to attribute this verse to the Synoptic accounts of the prayer in Gethsemane (*Hebrews 1-8*, 119-20), but early patristic sources make this connection. See, for example, Photius, *Fragments on the Epistle to the Hebrews* 5:7-9. Reference from Erik M. Heen and Philip D. W. Krey, eds., *Hebrews*, Ancient Christian Commentary on Scripture (Downers Grove: IVP Academic, 2005), 73-74.

On another level, as we noted earlier, when the Son identified with human beings by becoming human, he also assumed the stuff of Earth from which human beings were created (cf. Gen. 2:7). Viewed from this angle, the Son then suffers the frailty and weakness of a creation gone off the tracks. He is able to sympathize with a suffering creation and to minister on its behalf, effecting an eternal salvation for the whole Earth community. If the Son's priesthood is sufficient to attain salvation for human beings, then it is sufficient to attain it for all that the Son assumed when he became human.

However, having introduced the notion of a priesthood based on that of Melchizedek, the author pauses for the moment to exhort the readers on toward perfection before explaining in detail the substance of this new order of priesthood.

A Charge to Enter More Deeply into This Priesthood (5:11–6:12)

The final paragraph of ch. 5 at first glance appears odd, for having introduced the topic of the Melchizedekian priesthood, the author again avers from addressing it, this time to chide the audience for not being sufficiently mature to handle the apparently advanced nature of this teaching. Of course, the author will return to the topic after a brief interlude. Obviously the audience did not become miraculously mature in the time it took to hear ch. 6 read. The point here is to spur the audience on to maturity. Indeed, Heb. 6:1 gives voice to this intention, exhorting the pursuit of perfection by moving beyond the basic teachings of the faith. These teachings include repentance from dead works, faith toward God, baptism, laying on of hands, resurrection from the dead, and eternal judgment. It is not that these are unimportant, but that they are indeed already present among the community.

William Lane points out that each of the six elements listed corresponds to the development of the high priestly ministry that will occupy the author in 7:1–10:18.[3] As such, read ecologically, each of these doctrinal formulations has implications for understanding the vocation that the Son has reclaimed for Adam and his posterity. Ernst Conradie has argued that an ecological reading of the Bible must move beyond the interpretation of individual texts from the

---

[3] Lane, *Hebrews 1–8*, 140.

perspective of Earth to the articulation of the theological heritage of Christian faith in ecological terms.[4] Providing fully formed ecological interpretations of each of these doctrines exceeds the scope of the present study, but we may observe the function that the list performs at this point in the letter. Much of what we have observed to this point in our ecological reading points toward certain fundamentals of the faith that are foundational for critical engagement in environmental matters. Succinctly stated, the first three elements of the list—repentance, faith, baptism—speak to matters connected to participation in the new covenant. A fundamental reorientation to reality is implicit in these doctrines. In ecological terms, that reorientation involves seeing that the Son's ministry entails implications for the redemption of the whole cosmos, not just its human component. This understanding, as we have seen, is grounded in the doctrine of creation. That the doctrine of creation may be subsumed under the rubric of foundational teachings is illustrated in both the canonical ordering of the Hebrew Bible (cf. Genesis 1 and 2) and the starting point of the letter to the Hebrews itself (cf. Heb. 1:2-3). Creation is where it starts, and human relatedness within creation is quickly affirmed both in the early Genesis creation accounts (cf. Gen. 1:26-28; 2:15) and in Hebrews (cf. Heb. 2:6-8). The laying on of hands corresponds to the priestly roles with which both the first Adam and the Son were charged. The resurrection of the dead, as pertains to Jesus's resurrection, is that which enables the Son to destroy the power of death (cf. Heb. 2:14-15) and to lead many children to glory (cf. Heb. 2:10). As we noted in our discussion of Heb. 2:5-18 earlier, the attainment of glory is that which enables human beings to fulfill the vocation described in the author's quotation of Ps. 8:5-7 in Heb. 2:6-8. Finally, pertaining to the matter of eternal judgment, this would entail the consummation of the cosmos as the Son's inheritance (cf. Heb. 1:2). The list as presented here forces us to recall the ground we have covered already in our own ecological reading of the letter.

The author provides a cautious confirmation that indeed the discussion will move beyond what has already been established (Heb. 6:3). Then follows a stark warning about the severe consequences of failing to hold fast to the covenant in which the readers participate. Verses 4-6 are dire as they read,

---

[4] Ernst M. Conradie, "What on Earth Is an Ecological Hermeneutics? Some Broad Parameters," in *Ecological Hermeneutics: Biblical, Historical, and Theological Perspectives*, eds. David G. Horrell, Cherryl Hunt, Christopher Southgate, and Francesca Stavrakopoulou (London and New York: T & T Clark, 2010), 300–10.

intimating that were the readers to forsake their faith, given all that they have experienced, renewal unto repentance would be precluded. The list of the blessings in which the readers walk is quite impressive: they have been enlightened, tasted the heavenly gift, shared in the Holy Spirit, and tasted the goodness of the word of God and the powers of the age to come. The author is quick to affirm, however, that this warning is not actualized at present among his readers (6:9-12). The warning has its intended effect if it causes the readers to remain diligent in their covenantal faithfulness.

The effect of the warning viewed ecologically is essentially the same as the list of foundational teachings immediately preceding the warning. In effect, the author is saying that the community has been so blessed. What does this mean in an ecological reading? In effect, the community so blessed is living in the reality of the coming age even though that reality has not been consummated. If this is so, then we are to live in light of the eschatological destiny of creation. Are we functioning as priestly co-regents with God in creation, helping Earth to realize its divine destiny to become the dwelling place of God? An ecological reading of this warning forces us to stop here and dwell for a moment on what it is that a life so characterized implies about our place in the world. In effect, we may actually pause before proceeding on to ch. 7 for the more advanced teaching.

One last observation is noteworthy here. It is interesting that the author here chooses to cement the warning with illustrations drawn from the natural world. In vv. 7-8, the author uses the images of ground, botanical produce, and rain to illustrate the effect of the warning. The whole dynamic demonstrates the principle of growth. God's blessings, the rain, shower down on all. It simply matters what kinds of seeds are being so watered. The author is convinced the audience consists of good seeds that will thrive with God's blessings. However, the rain may only produce thorns and eventuate in a curse. The image draws upon God's judgment on the first humans in the Garden of Eden (Gen. 3:17-18). I have written elsewhere on the dynamic of the divine/human/other-than-human triad in which God and the other-than-human order cooperate either to bless or to judge human beings.[5] Read ecologically, the imagery here brings to mind this dynamic in which the harmonious function within the triad is interrupted and in which God and the other-than-human order work through

---

[5] Lamp, *Reading Green*, ch. 5.

both positive and negative effects to bring about human cooperation so that all creation might thrive as intended.

Abraham and God's Covenantal Faithfulness (6:13-20)

For the first time in the epistle, Abraham is mentioned. For the author of Hebrews, Abraham is a figure worth engaging, for Abraham is noteworthy precisely in connection with the concept of covenant. At this stage, however, the issue is not to compare Abraham and Jesus, but to show, via Abraham, that God is faithful to the covenants God makes. When God made the covenant with Abraham, God is shown as first having made promises to Abraham, and then sealing these promises with an oath. Since God proved faithful in fulfilling the particularly mentioned promise—"I will surely bless you and multiply you" (6:14)—God may be trusted to fulfill the covenantal promise through Jesus. Here the promise is framed in terms of the hope that awaits those who remain faithful. Yet in a wonderful segue to the long awaited discussion of the Son's Melchizedekian priesthood, the author characterizes this hope as one that emboldens us to enter the inner shrine, where Jesus has gone before us as a high priest of the order of Melchizedek.

Abraham is a crucial figure in an ecological reading of Hebrews. Indeed, the author of Hebrews has provided the very point of connection for an ecological reading in the quotation of Gen. 22:17 in Heb. 6:14: "I will surely bless you and multiply you." To see this, we turn to N. T. Wright's magisterial study of Paul, in which Wright identifies a trajectory of the conception of the people of God that extends from the first pages of the Old Testament into early Christianity.[6] According to Wright, "As far as Paul was concerned, the reason the creator God called Abraham in the first place was to undo the sin of Adam and its effects."[7] The promises made to Abraham echo the commands given to Adam in Gen. 1:28. Adam was to be fruitful, multiply, and fill the earth. Abraham was promised that he would be the father of a great nation and inhabit a land (Gen. 12:2-3; 17:2, 6, 8; 22:16-18). This pattern of promise was repeated to Isaac (Gen. 26:3-4, 24) and to Jacob (Gen. 35:11-12), as well as in Isaac's blessing of

---

[6] Wright, *Paul and the Faithfulness of God*, 783–95. Wright's study is more conceptually oriented toward a discussion of Paul's view of ecclesiology, and its citation here should not be seen as indicating that Wright attributes the authorship of Hebrews to Paul.
[7] Wright, *Paul and the Faithfulness of God*, 784.

Jacob (Gen. 28:3-4) and in Jacob's blessing of Joseph (Gen. 48:3-4). Abraham and his descendants are depicted in parallel with Adam's vocational mandate.

But for Wright, this is more than mere parallel presentation. Rather, it is a sequence extending from Adam to Abraham[8] and then forward ultimately to Christ. The connection of Abraham's calling to Adam's is apparent in two senses. First, Abraham is connected to Adam in terms of creation. Adam was called to multiply and fill the earth. Abraham was called to this as well. So in terms of the vocational mandate for Adam to fill creation with his offspring, Abraham is charged to recapitulate this mandate. Here the calls are paralleled. Second, Abraham is connected to Adam in terms of Adam's fall. In Adam, death and corruption enter the world; through Abraham, all nations of the world will be blessed. Here, Abraham's calling is to rescue fallen Adam.[9] "If Abraham and his family thus *recapitulate* the role of Adam, they are also the ones in whom the creator God determines to *rescue* the human race from its plight."[10] So the connection between Adam and Abraham is both resumptive and redemptive of the human vocation.[11]

Of course, much theological reflection on this trajectory is spent on how this effects the redemption of human beings. Anthropocentric bias is evident here. But if Abraham is shown to be the recapitulation of Adam, then it stands to reason this should also include the Adamic vocation of which we have spoken so frequently. Abraham does not simply help provide for the redemption of Adam's posterity, but he also goes about all that Adam was to do in the world. God's reclamation project extends beyond human beings to include all over which Adam was to exercise benevolent priestly co-regency with God. It is in effect the antidote to the disruption introduced by Adam into the God/human/other-than-human triad.

In the context of Heb. 6:13-20, mention of Abraham forms a bridge to the discussion of the priesthood of the Son in the order of Melchizedek. In an ecological reading, mention of Abraham provides another piece of evidence that God's plan for the ages was always to rule creation with God's image-bearing priestly co-regents. It was Adam's vocation, and it was integral to the Abrahamic covenant that is fulfilled in the Son. As such, Abraham's mention

---

[8] Prior to Abraham, a similar command is given to Noah in Gen. 9:1.
[9] Wright, *Paul and the Faithfulness of God*, 784–85.
[10] Wright, *Paul and the Faithfulness of God*, 787.
[11] Wright, *Paul and the Faithfulness of God*, 788.

here fits well as a prelude to the discussion of the Son's priesthood. On the one hand, as we have argued, Abraham fits within the trajectory of the recapitulated Adamic vocation. On the other hand, Abraham is a crucial figure in the upcoming discussion of Melchizedek. In ch. 7, the author will finally address the full significance of Melchizedek after a few earlier teases.

Hearing Earth

Perhaps Earth's call here is for human beings to take seriously a thoroughgoing articulation of the Christian theological tradition in ecological terms. For too long, the primary focus of Christian theology has been anthropocentrically driven. In the present era, however, human beings have the capacity to wreak havoc on the biosphere on a global scale. So substantive is the effect of human activity on the planet and its environment that many are calling for the formal adoption of a new geological age to reflect this impact. Labeled the "Anthropocene" by many, the term reflects a shift in the relationship between human beings and their world, from a time when human beings were primarily shaped by their environments to a time now where human beings significantly affect the environment.[12] Such a fundamental shift requires a new focus in theology, one that directs attention to the God/human/other-than-human triad.

As mentioned earlier, Ernst Conradie has issued such a challenge. The major categories of systematic theology need a rearticulation in ecological terms, or perhaps rather, a rigorous constructive ecological theology needs development.[13] Many preliminary steps have been taken, consisting of treatments advocating for the care of creation across the theological spectrum.[14] Important as these studies are, they all tend to focus on ecological

---

[12] For a brief discussion, see Lucy E. Edwards, "What Is the Anthropocene?" *EOS: Earth and Space Science News*, November 30, 2015, https://eos.org/opinions/what-is-the-anthropocene (accessed July 15, 2019).

[13] As noted in the introduction of this volume, the primary difference between Norman Habel's model of ecological hermeneutics and that of the Exeter Project is the latter's advocacy for such a theological project. For a discussion, see Lamp, *Reading Green*, 11–16.

[14] For example, Roman Catholic: Pope Francis, *Laudato Si': On Care for Our Common Home* (Vatican City: Libreria Editrice Vaticana, 2015); Eastern Orthodox: John Chryssavgis and Bruce V. Foltz, eds., *Toward an Ecology of Transfiguration: Orthodox Christian Perspectives on Environment, Nature, and Creation* (New York: Fordham University Press, 2013); Bartholomew I, *On Earth as in Heaven: Ecological Vision and Initiatives of Ecumenical Patriarch Bartholomew*, ed. John Chryssavgis (New York: Fordham University Press, 2012); Protestant: Jürgen Moltmann, *God in Creation: A New Theology of Creation and the Spirit of God*, trans. Margaret Kohl (San Francisco: Harper & Row,

theology, or ecotheology, as a subdiscipline within theology. What is needed, rather, is a theological treatment that takes the cardinal doctrines and dogmas of the Christian tradition and rearticulates them ecologically.[15] Only when ecological concerns are addressed within the very structure of Christian theology will attempts at constructing ethical paradigms and plans of action attain the necessary gravitas to become essentials of the faith rather than mere accoutrements.

1985); Evangelical: Van Dyke, *Between Heaven and Earth*; Ben Lowe, *Green Revolution: Coming Together to Care for Creation* (Downers Grove: IVP, 2009); Richard Bauckham, *The Bible and Ecology: Rediscovering the Community of Creation* (Waco: Baylor University Press, 2010); Howard A. Snyder and Joel Scandrett, *Salvation Means Creation Healed: The Ecology of Sin and Grace* (Eugene: Cascade, 2011); Pentecostal: Swoboda, *Tongues and Trees: Toward a Pentecostal Ecological Theology*; A. J. Swoboda, ed., *Blood Cries Out: Pentecostals, Ecology, and the Groans of Creation* (Eugene: Pickwick, 2014); Feminist: Sally McFague, *The Body of God: An Ecological Theology* (Minneapolis: Augsburg Fortress, 1993).

[15] I have begun a monograph to do this very thing, and it has been the focus of several paper presentations in recent years at the Society of Biblical Literature annual meetings. At this writing, the volume is approximately half written.

7

Jesus: The New Adamic Priest (Heb. 7:1-28)

Contents

The author now comes to a significant moment in the argument, where the priesthood of Jesus is established as a qualitatively different and superior priesthood over against the Levitical priesthood of the old covenant. After some linguistic and genealogical gymnastics, the author enters into a detailed explanation of the nature of the two priesthoods that makes Jesus's priesthood superior. This step is foundational for the arguments of the next two and one-half chapters, where aspects of the old covenant connected with priesthood are compared unfavorably to their new covenant counterparts in Jesus. From an ecological perspective, the passage describes a priesthood that is integrally connected with the covenant made with Abraham in the process of reclaiming the Adamic vocation for human beings. Priesthood is a key image for the Adamic vocation as described in Genesis 1 and 2. The depiction of Jesus as a priest of an eternal order fulfills the proleptic role of the Levitical priesthood as well as reestablishes the priestly vocation of the first Adam, which was to be an eternal priesthood in its own right. Now that the vocation has been reclaimed, it may entrusted to those who participate in the new covenant.

Structure

7:1-10: Jesus and Melchizedek: the Abraham connection
7:11-28: The new Adamic Priest

Analysis

Jesus and Melchizedek: The Abraham Connection (7:1-10)

With ch. 7, the author of Hebrews finally embarks upon the discussion that has been promised from as early as Heb. 2:17: the discussion of the high priesthood of Jesus. In the discussion of 4:14–5:10 this priesthood is connected with the priesthood of one of the most enigmatic figures in the Old Testament: Melchizedek. Now in ch. 7, the figure of Melchizedek is mined for his contribution to the Son's priesthood.[1]

For one who figures so prominently in the author's discussion of Jesus's priesthood, Melchizedek is remarkably mentioned in only two passages in the Old Testament. Of special interest in Hebrews is Ps. 110:4, which is quoted in Heb. 5:6 and 7:17. This single verse contains both features of Melchizedek's priesthood that establish the foundation of the Son's priesthood. First, the mention of "forever" (εἰς τὸν αἰῶνα) is used to establish the eternality of the Son's priesthood in contrast with the temporary nature of the priests descended from the line of Levi. This will receive a rather creative exegetical justification in the opening paragraph of ch. 7. Second, Melchizedek represents a whole new order of priesthood separate from the Levitical priesthood, which justifies the Son's occupancy of a priestly order outside of the Levitical order.

The other Old Testament passage in which Melchizedek appears is in a brief narrative in Genesis 14, where he appears to a victorious Abram who had just defeated a retinue of kings in the rescue of his nephew Lot. The narrative in Gen. 14:18-20 reads as follows:

> And King Melchizedek of Salem brought out bread and wine; he was priest of God Most High. He blessed him and said,
>
> > "Blessed be Abram by God Most High,
> > > maker of heaven and earth;
> >
> > and blessed be God Most High,
> > > who has delivered your enemies into your hand!"
>
> And Abram gave him one tenth of everything.

---

[1] I discuss this passage in *The Greening of Hebrews?*, 92–95.

The author of Hebrews is, however, quite selective in which parts of the narrative are appropriated in the discussion of Jesus's priesthood.

The author begins with a brief series of etymological exercises that implicitly connect Melchizedek's name and geographical location with the Son. The name Melchizedek derives from two Hebrew words that mean "king" (מלך) and "righteousness" (צדק), so he is identified as the "king of righteousness" (Heb. 7:2). Next, since Melchizedek is also the king of Salem (שלם), this establishes him as "king of peace" as well. Interestingly, the author of Hebrews does not make much of these points, except simply to state them. However, we may draw upon them in an ecological reading, which we will do shortly.

The priestly credentials of Melchizedek derive from his identification as "priest of the Most High God" (7:1; cf. Gen. 14:18). As such, he blessed Abraham, affording the author of Hebrews the opportunity to identify a high priestly lineage that does not derive from Levi but is superior to Levi. This is the gist of Heb. 7:4-10. The hermeneutical key is in v. 7: inferior parties are blessed by superior parties. Since Melchizedek blessed Abraham, Melchizedek is superior not only to Abraham but also to the entirety of Abraham's lineage, including Levi. Priests collect tithes from their own people; Melchizedek collected tithes from Abraham, who is not connected this way with Melchizedek. So in effect, Melchizedek collected tithes from Levi, and thus is superior to Levi. At this point, the author of Hebrews has established a taxonomic framework for demonstrating the superiority of Melchizedek's priesthood over the Levitical priesthood.

Hebrews 7:3 establishes the more important characteristic of Melchizedek's priesthood as pertains to the Son's priesthood. Here, the author draws on certain literary features of Gen. 14:18-20 to argue that Melchizedek's priesthood is of an eternal order. There is no genealogical information given of Melchizedek; no father or mother is mentioned. Moreover, there is no mention of his death. From these two omissions the author of Hebrews concludes that Melchizedek, without beginning or end of life, resembles the Son of God, who is eternal. More importantly, the priesthood of Melchizedek is eternal.

Of course, the interests of the author of Hebrews here are christological. Hebrews 7:1-10 is the foundation upon which the high priesthood of Jesus will be defended in the remainder of the chapter. But in this odd section, there are features that lend themselves to ecological reflection. Here, we will adopt a subversive tack in reading Gen. 14:18-20, focusing not so much on what the author has cited but on what was omitted. In the same way the author exploited omissions in Gen. 14:18-20, we will exploit omissions in this passage. First, the author does not cite the second clause of Melchizedek's blessing of Abraham in which Melchizedek identifies God as "maker of heaven and earth." As we have noted on several occasions, the doctrine of creation figures significantly into the author's argument. Most pertinently, it is through the Son that God created the heavens and earth (cf. Heb. 1:2). Omission of this attribution may be seen as a bias in the author's argument, for it takes attention away from Earth and serves the particular christological interests of the author. However, recovering this attribution here enables a more fully orbed understanding of the Son's priesthood as it is described in 7:11-28. The Son's priesthood must consider the Son's role in creation as well as his assumption of the stuff of creation in his incarnation. Here Melchizedek, in effect, recalls the creative work of God, which in Hebrews is achieved through the Son, and locates it under the rubric of the Son's priesthood.

The more significant omission from Gen. 14:18-20, one that has great implications for our ecological reading, is the omission of the reference to Melchizedek serving bread and wine to Abraham. From very early times, Christian thinkers have seen in Gen. 14:18 an allusion to the sacrament of the Eucharist.[2] The omission of the reference to bread and wine in Hebrews has given rise to speculation on the role of the sacrament in the community addressed by the letter, ranging from identification of oblique references to the sacrament in passages in the letter (e.g., Heb. 13:10), to suggestion that the sacrament itself forms the infrastructure of the entire letter, to assertion that Christ has rendered all ritual practice of any kind obsolete.[3] The simplest solution to the omission of the phrase by the author of Hebrews was that it did not serve any purpose in the author's argument. In terms of ecological hermeneutics, its omission evidences a bias on the part of the author. Again, our subversive tack here is to reappropriate the omitted language and plumb it for its ecological significance.

---

[2] For example, Eusebius of Caesarea, *Proof of the Gospel* 5:3; Cyprian, *Letter* 62:4; Epiphanus of Salamis, *Panarion* 4, *Against Melchizedekians* 6:1-11; Clement of Alexandria, *Stromateis* 4:25; Jerome, *Hebrew Questions on Genesis* 14:18-19. References from Heen and Krey, eds., *Hebrews*, 95–96, 97–98, 99–100, 102, 104, respectively.

[3] R. Williamson, "The Eucharist and the Epistle to the Hebrews," *New Testament Studies* 21 (1975): 300–12.

In my previous volume, I discussed how the Eucharist might contribute to an ecological reading of the passage, though my point of departure there was Heb. 13:10: "We have an altar from which those who officiate in the tent have no right to eat."[4] In that study, the Eucharist functioned as the point of identification from which the voice of Earth might be retrieved. The work of Roman Catholic theologian Denis Edwards provided the substance of the reflections in that study.[5] For the present study, we will take a different approach to the issue. Here, our concern is with the author's discussion of priesthood, for it is within the context of this discussion that the author of Hebrews invokes the Melchizedek tradition. As we have noted several times, the Adamic vocation of human beings was to be the priestly co-regents with God in the world. For the majority of Christians throughout the history of the church, the Eucharist has been the central act of corporate worship, an act performed by those functioning in an official priestly role. Our concern here will be to understand how the Eucharist may help in the restoration of the Adamic vocation to human beings through the high priesthood of the Second Adam. This time, I will look to the Eucharistic practice of Eastern Orthodoxy, especially as expressed in an enigmatic line in the Divine Liturgy of St. John Chrysostom, to frame our discussion.

In the Divine Liturgy, the central act is the priestly offering, or anaphora, of the bread and wine to God on behalf of the people. In this part of the Liturgy, the priest speaks aloud the line, Τὰ σὰ ἐκ τῶν σῶν σοὶ προσφέρομεν κατὰ πάντα καὶ διὰ πάντα.[6] The first part of the line—Τὰ σὰ ἐκ τῶν σῶν σοὶ προσφέρομεν—is not difficult to translate: "Offering to you your own of your own." It is the remainder of the line—κατὰ πάντα καὶ διὰ πάντα—that presents difficulty. Translated quite woodenly, it might read, "according to all and for all." Without getting overly bogged down in the technical details,[7] what provides the greatest difficulty in the second part of the line is that the word πάντα, "all," which we have encountered previously, may actually be either a masculine singular or a

---

[4] Lamp, *Greening of Hebrews?*, 85–100.
[5] Denis Edwards, *Ecology at the Heart of Faith* (Maryknoll: Orbis, 2006).
[6] John Chrysostom, *The Divine Liturgy of Our Father among the Saints, John Chrysostom*, trans. Seraphim Dedes (Columbia: Newrome Press, 1995), 60.
[7] For a discussion of the various interpretive problems, see Andreas Andreopoulos, "'All in All' in the Byzantine Anaphora and the Eschatological Mystagogy of Maximos the Confessor," in *Studia Patristica*, vol. 68, ed. Markus Vinzent (Leuven: Peeters, 2013), 303–12; Emmanuel Hatzidakis, *The Heavenly Banquet: Understanding the Divine Liturgy* (Chicago: Orthodox Witness, 2010), 255–59.

neuter plural form of the word. Unlike the biblical references that we examined earlier, there is nothing in the context in the Divine Liturgy to indicate the gender of either occurrence. Moreover, there is scant comment on this line in the various commentators on the Divine Liturgy throughout history, and the context of the occurrence of the line in the Liturgy does not narrow down the possibilities. The resulting state of affairs is that any attempt at identifying the exact meaning of the phrase is nothing more than "educated, even inspired guesswork."[8]

This does not lead to a counsel of despair. We might simply read the line from an ecological perspective, as we are doing with the letter to the Hebrews. Indeed, some within the Orthodox tradition today are doing this. Though not explicitly addressing the matter from an ecological perspective, Alexander Schmemann nevertheless reads the Eucharist from the perspective of the Adamic vocation. The primary definition of being human is that human beings are the priests of the world, standing in the center of creation and unifying it in an act of blessing God, "of both receiving the world from God and offering it to God—and by filling the world with this eucharist. . . . The world was created as the 'matter,' the material of one all-embracing eucharist, and [human beings were] created as the priest[s] of this cosmic sacrament."[9] In the original Adamic vocation, human beings were priests offering up creation itself to God. However, given the reality of the fall and the subsequent reclamation of the Adamic vocation in Christ, now the sacrament of the Eucharist becomes "our offering to [God] of ourselves, of our life and of the whole world."[10] Indeed, "in and through this Eucharist the whole creation becomes what it always was to be and yet failed to be."[11]

Of course, the actual mechanics of the Divine Liturgy in the line in question state that we offer up to God that which is always God's. It is no accident that what is offered is bread and wine, those things that are the products of human industry made from that which is divine gift to us from Earth. Moreover, in the Divine Liturgy, the priest calls upon the Holy Spirit to make the gifts of bread and wine the very body and blood of Christ. In his incarnation, as we

---

[8] Andreopoulos, "All in All," 308.
[9] Alexander Schmemann, *For the Life of the World: Sacraments and Orthodoxy* (Crestwood: St. Vladimir's Seminary Press, 1973), 15.
[10] Schmemann, *For the Life of the World*, 35.
[11] Schmemann, *For the Life of the World*, 38.

have noted, the enfleshed Son takes to himself, to his very bodily constitution in which inhabits the fullness of deity (cf. Col. 2:9), the stuff of Earth. In the act of the mysterious transformation of food items into the body and blood of Christ, Christ identifies with the material creation, and in the offering of this bread and wine to God, human beings function in their divinely ordained priestly vocation to restore the God/human/other-than-human triad to communion.[12] Moreover, since human beings, by their very nature, are food-consuming creatures, by partaking of the bread and wine they in turn receive into their very lives the sustaining gifts not only of Christ but also of Earth.[13] In the sacrament of the Eucharist, as understood by Eastern Orthodoxy, there is the potential for an ecological appropriation of the sacrament that informs the Adamic vocation that has been reclaimed in the priesthood of the Son.[14]

In terms of Heb. 7:1-10, what we have done here is to subvert the author's rhetorical strategy by drawing into discussion that which was omitted. Both the mention of God as creator of heaven and earth and Melchizedek's offering of bread and wine were omitted by our author, thus cementing an anthropocentric bias in the reading of the passage. However, once we draw in the whole account of Melchizedek from Gen. 14:18-20, we find points of identification with Earth that give further definition to the Melchizedekian priesthood that forms the template for the Son's own high priesthood. The remainder of Hebrews 7 is the defense of the author's claim that Jesus is the one who fulfills the promise of a new and better priesthood.

The New Adamic Priest (7:11-28)

Though this section of ch. 7 is lengthier than the previous section, its place in the argument is rather simple and straightforward. Hebrews 7:1-10 established a template for the new, better high priest. Hebrews 7:11-28 effectively argues that Jesus is the one who follows in the pattern of the priesthood of Melchizedek. Since Jesus is not descended from the proper tribe to function as a high priest, a new order of priesthood becomes necessary (7:13-14), and when such a

---

[12] John Zizioulas, "Preserving God's Creation: Three Lectures on Ecology and Theology," *King's Theological Review* 12 (1989): 1, 4–5.
[13] Schmemann, *For the Life of the World*, 42–43.
[14] See also George Theokritoff, "The Cosmology of the Eucharist," in *Toward an Ecology of Transfiguration: Orthodox Christian Perspectives on Environment, Nature, and Creation*, eds. John Chryssavgis and Bruce V. Foltz (New York: Fordham University Press, 2013), 131–35.

priesthood is established, a change in the law is necessary (7:12). Interestingly, though, the author of Hebrews asserts that the new priesthood on the order of Melchizedek is not established on a new, better law but on the basis of a new hope (7:19). We will pursue this matter in our discussion of ch. 8.

Moreover, the author makes a special point about the inability of the law to make purification of sins. The law, the author asserts, was unable to make anything perfect (7:19). Indeed, the point was made earlier that the human high priests first had to make purification for their own sins before offering sacrifices for the purification of the people (cf. 5:3). Indeed, Jesus the new high priest has no need to offer sacrifices for his own sins because he is without sin (7:26-28). Later in the epistle, the sacrifice that Jesus offered will purify the people so that they will be able to worship God with a purified conscience (9:13-14). In the logic of the argument, the perfection of Jesus, evidenced in the power of his indestructible life (7:16), is the basis of that which allows human beings to worship God rightly. If, as we have argued, a significant aspect of Jesus's redemption involves the reclamation of the Adamic vocation for human beings, then it is not a huge leap to argue that the ability to worship God from a purified conscience is a crucial element of the ability of human beings to function in priestly co-regency with God in creation. Though such does not seem to be preeminent in the mind of the author of Hebrews, viewed from an ecological perspective, the notion that human beings are to follow in the priesthood exercised by Jesus is easily derived from further consideration of those things argued by the author of Hebrews. The New Testament elsewhere characterizes redeemed human beings as a nation of priests and kings, which fits precisely with the depiction of the Adamic vocation of Gen. 1:26-28 (e.g., 1 Pet. 2:5, 9; Rev. 1:6; 5:10). The author of Hebrews does not tease this out because this is not the thrust of the argument of the epistle, nor does it serve the purpose of exhortation prevalent in the letter. However, the notion that the priesthood of Jesus serves as a model of the priesthood exercised by human beings is not far away, especially when the author's argument is viewed from an ecological perspective.

Circling back to pick up the omitted reference to bread and wine in Gen. 14:18, the Eucharist becomes the prototypical liturgical act that expresses the reclaimed Adamic vocation for human beings. In the Divine Liturgy, as the gifts of bread and wine are brought forward for offering, the priest prays in words reminiscent of the letter to the Hebrews: "Look upon me, your sinful and unprofitable servant, and purify my soul and heart from an evil conscience."[15] Here the impulse to view human priestly service in the offering of the Eucharist is derived from Christ's high priesthood. Jesus, the divine, eternal Son, who is established as a high priest in the order of Melchizedek, purifies human beings so that they might operate in their Adamic priesthood in the world. Moreover, in the Divine Liturgy, just prior to the line that occupied our attention in the previous section of this discussion, the priest prays, "We thank you also for this liturgy which you have been pleased to accept from our hands, though there stand around you thousands of archangels and tens of thousands of angels, the Cherubim and the Seraphim, six-winged and many-eyed, soaring aloft on their wings."[16] Here acknowledgment is made that, despite the fact that God is surrounded by creatures of heavenly splendor, it is appropriate that the Eucharistic offering be made by human beings. In light of our ecological approach, this is the case precisely because it accords well with the Adamic vocation to be God's priestly co-regents in the world. Of course, it would be folly to ascribe to St. John Chrysostom an ecological motivation for composing this prayer. But in light of today's ecological crises, this prayer aligns with a reading of the human vocation in the world as one in which human beings are charged with the care and thriving of Earth. Eucharist is that priestly act that unites the people with Christ, offering themselves with Christ to God in the bread and wine and offering the world back to God who in turn blesses human beings with the world so that human beings might become a blessing to the world.

Eucharist becomes the link between Jesus, the high priest according to the order of Melchizedek, and human beings who accept from the ascended Jesus the mantle of this priesthood so that they might operate in their original Adamic vocation. This link is only possible because the author of Hebrews has provided for a subversive reading of Gen. 14:18-20 by appealing to Melchizedek as the template of the Son's eternal high priesthood. Though we have focused in this discussion on an Eastern Orthodox perspective of the Eucharist, this tack would accord well with any tradition that holds a sufficiently high sacramental theology of the Eucharist.[17]

---

[15] Chrysostom, *Divine Liturgy*, 50.
[16] Chrysostom, *Divine Liturgy*, 58.
[17] We have already noted that my previous volume focused on the thought of a Roman Catholic theologian, Denis Edwards. But some intriguing work is being done in Pentecostal circles by Chris

In the next chapter of the letter, the author draws a conclusion from the discussion on the Son's priesthood as pertains both to other aspects of the old covenant cultus and the law. This will open up other vistas for ecological implications of the Son's new covenant for Earth.

Hearing Earth

In light of this discussion, how might Earth speak to those who, by virtue of their adherence to the new covenant instituted by the Son, have had their Adamic vocation reclaimed for them by the Son? Here we might see Earth engaged in a similar exercise as the author of Hebrews early in Hebrews 7: appeal to the appellations attributed to Melchizedek. In Heb. 7:2, the author parses Melchizedek's name to show that his name means he is "king of righteousness" and that his designation as king of Salem means he is the "king of peace." As we noted, once stated, these data go unexplained by the author of Hebrews. Earth would return attention to these facts and bring them into conversation with the Adamic vocation to serve as co-regents with God in creation.

In the LXX the Greek term translated "righteousness," δικαιοσύνη, renders the Hebrew term צֶדֶק, which means acting according to God's standard or divine requirement, in accordance with covenant fidelity.[18] Earth might also remind us that in the LXX, the term δικαιοσύνη also renders the Hebrew term חֶסֶד, referring to kindness, mercy, and loyalty.[19] Bringing these two ideas together, Earth might suggest that in the name Melchizedek, what we see is actually a priest-king who exercises God's righteousness on behalf of Earth in fulfillment of the Adamic vocation regained by Christ and bequeathed to those who, after him, function in this vocation. In the Melchizedekian priesthood of Christ, human beings should establish covenantal righteousness on behalf of Earth, bringing together ideas of right action and mercy to help Earth achieve its destiny to become the dwelling place of God. A similar move occurs with

---

E. W. Green, who argues for the reclamation of a high sacramental view of the Eucharist that opens possibilities for similar ecological applications. See his volume, *Toward a Pentecostal Theology of the Lord's Supper: Foretasting the Kingdom* (Cleveland: CPT Press, 2012). Similar liturgical resources are available in Norman Habel, David Rhoads, and H. Paul Santmire, *The Season of Creation: A Preaching Commentary* (Minneapolis: Fortress, 2011).

[18] *NIDNTTE*, 1:725.
[19] *NIDNTTE*, 1:725.

the attribution of Melchizedek as the "king of peace." In the LXX the Greek term εἰρήνη almost always renders the Hebrew word שָׁלוֹם, which conveys a sense of completeness, wholeness, or perfection.[20] Earth might remind human beings that to function in the Melchizedekian priesthood of Christ is to function as one who restores wholeness to Earth as priestly co-regents with God.

One last lexical lesson Earth might give to human beings regards the omitted detail from the Melchizedek narrative of Gen. 14:18-20: the bread and wine. Our ecological reading recovered this detail and framed it in terms of the sacrament of the Eucharist. Earth would remind us that the word "Eucharist" derives from the Greek word εὐχαριστία, which means thanksgiving.[21] Earth would call human beings to regain a sense of thankfulness for all that Earth provides for human thriving, and to ask for commensurate responses of gratitude on behalf of Earth. Partaking of the sacrament of the Eucharist, on the one hand, is an expression of gratitude for all that God has done for us in Christ; on the other hand, in the offering of the fruits of Earth in the bread and wine, human beings receive back from God the sustaining blessings of Earth that not only convey spiritual blessings but also remind us of all that Earth provides for human thriving in the world.

The figure of Melchizedek functions not only as a template of the high priesthood of Jesus but also as the model of how human beings should function as heirs of the reclaimed Adamic vocation in the Son. Earth effectively allocates the high priesthood of the Son to the spiritual descendants of the Son, those who are to function as the royal priesthood in the world. While our discussion above regarding the Eucharist focused heavily on the liturgical order of the Eucharist in Eastern Orthodox tradition, Eucharist is not simply the purview of the ecclesiastical orders of ministry. Rather, those who function liturgically as priests do so to equip the people to be people of Eucharistic piety in the world. Earth simply reminds us that the world consists of more than just human beings. We are to carry the blessings of the Eucharist back to Earth who provides us with the raw materials for our celebration of the sacrament so that we might in turn convey the blessings of divine communion with Earth.

---

[20] *NIDNTTE*, 2:112.
[21] *NIDNTTE*, 2:334.

8

A Logic of the New Order (Heb. 8:1-13)

Contents

If, as the author of Hebrews has argued, Jesus is the eternal, perfect high priest of the new covenant, then this reconfigures everything with which the priest is connected. There must be a new way of conceiving of the sacrifices offered by this priest as well as a new way of thinking of the place those sacrifices are offered. Indeed, since this whole apparatus is so integral to the old covenant, to change it entails the establishment of a whole new covenant. This chapter, then, functions to make this claim and to move the discussion forward in terms of priestly performance. Ecologically speaking, this chapter functions in the same way, though by looking at the priesthood of Jesus in Adamic categories. Here we begin to view how such things as sacrifices and the sanctuary function ecologically. Indeed, we must assess how the new covenant is understood ecologically, including even the law.

Structure

8:1-7: A whole new priestly context
8:8-13: Reclaiming the law for Earth

Analysis

A Whole New Priestly Context (8:1-7)

Chapter 8 functions, for the author of Hebrews, as the fulcrum of the letter. Occurring near the center of the argument, it both draws a conclusion from the preceding argumentation in which Jesus is established as the high priest along the order of Melchizedek and begins to move the argument forward by showing how there must be commensurate offerings for such a perfect, eternal high priest as well as an appropriate place of offering. The matter of a sanctuary arises first. If our high priest is seated at the "right hand of the throne of the Majesty in the heavens" (8:1), then there must be a sanctuary for his priestly service worthy of such an exalted high priest. At this point, the author of Hebrews asserts that such a sanctuary cannot be the one on earth, the one Moses was commanded to build, for such a tent would have been erected by the hands of mere mortals. For this high priest, there must be a tent established by the Lord (v. 2). Moses was commanded to build a tent for worship that was patterned after the heavenly sanctuary he was granted to see while on the mountain (8:5). Despite the heavenly origin of this vision, the resulting product is characterized as a mere "sketch" and "shadow" of the heavenly sanctuary in which Jesus offered his sacrifice.

As the author has asserted in ch. 7, in terms of earthly credentials, Jesus would not even be qualified to be numbered among the high priests, for he was not descended from priestly lineage. Given this, he would not be able to offer the sacrifices of these priests. Therefore, he must have an appropriate offering to make. The author does not develop this line of thinking at this point; it is merely stated as a corollary to the necessity of a new sanctuary. A whole new priestly apparatus is envisioned here. A new priest, indeed a whole new kind of priest, is necessary to deal with the human condition. If such a new priest is required, then all that goes into the performance of priestly ministry must be reconfigured as well. This is not simply a new order of ministry; it is also a superior one. Indeed, all of this is now framed in the necessity of a completely new covenantal relationship between God and the people. The old one simply cannot stand, for it was wholly inadequate to deal with the sins that separated the people from God. This "more excellent" ministry is grounded in better promises, and as such needs a new covenantal framework.

This section may be seen, ecologically speaking, as biased against Earth. Here we see language that hints at a sort of dualism that views heavenly realities as the ultimately real over against the material realities that are, at best, merely patterned on the heavenly realities and are destined to pass away when the new has come (cf. 8:13). Nevertheless, we cannot so cavalierly pass over the value of the material realities. While they may be mere copies of heavenly realities, God deigned to be known in and through them. So while they may have been crafted by human hands, they were still avenues through which God was made known to and experienced by human beings.

More pertinent from an ecological perspective is the language of v. 2 that this is "the sanctuary and the true tent that the Lord, and not any mortal, has set up." Crucial is the verb πήγνυμι, which frequently carries the sense of construction or building.[1] It seems unlikely that the verb here intimates that in the heavenly realm there exists a sanctuary that is built by God and serves as an architectural blueprint for the one Moses built on earth. The assertion here does not seem to be a metaphysical pronouncement but is rather a temporal pronouncement comparing the natures of the worship carried out in the old and new covenants.[2] This clearly serves the purpose of the author of Hebrews. So how might a verb connected with construction of a heavenly sanctuary inform an ecological reading of this passage?

Here we revisit what we have asserted throughout our study, namely, that the Adamic vocation entailed human beings serving as priestly co-regents with God to help creation realize its destiny to become the dwelling place of God. Whatever the liturgical function the heavenly sanctuary has in the economy of God's redemption in Christ, we cannot lose sight of the fact that God's creational intent seems to entail the desire to dwell with all that God has created. Surely the created cosmos would satisfy the definition of being a dwelling that was not set up by human beings. Of course, this is not what the author of Hebrews had in mind. But language such as "tent" is used to indicate dwelling as well as a worship setting. Indeed, as far back as 3:1-6 this line of thinking emerges. There Moses was a faithful servant in God's house. Every house, the author asserts, is built by someone, but God is the builder of all things (v. 4). Indeed, in v. 6, the faithful people constitute the house over which Christ is the Son. This seems to suggest that in Hebrews, the idea of God's dwelling place is flexible, capable of various modes of expression. Perhaps here in ch. 8, both the earthly dwelling constructed by Moses on the model of the heavenly sanctuary and the heavenly sanctuary itself are anticipatory of the time when the whole of creation, a dwelling not made with mortal hands, becomes God's dwelling place.

---

[1] BDAG, 811.
[2] Lane, *Hebrews 1–8*, 206.

Reclaiming the Law for Earth (8:8-13)

In the previous chapter, the author of Hebrews presented the argument that when a change in priesthood occurs, there was necessarily also a concordant change in law (7:12). Moreover, the author also asserted that the law had been abrogated due to its weakness and ineffectiveness to deal with sin (v. 18), for "the law made nothing perfect" (v. 19). Interestingly, though, this notion of a change in law occasioned by the change in priesthood, is not immediately developed; no indication is given of what this new "law" consists. Instead, the author argued that this new priesthood introduced a "better hope" (v. 19), for this eternal priesthood is confirmed with God's oath (v. 21; cf. Ps. 110:4). As we saw in the opening section of Hebrews 8, Jesus's ministry is of a "more excellent" quality that is enacted through "better promises" (8:6). Indeed, the better covenant is necessary because the first one was ineffectual. Yet as of v. 7, there is no development of the new law that comes in alongside the new priesthood.

With v. 8, however, the concept of law reemerges with a lengthy quotation of Jer. 31:31-34. For the author of Hebrews, the citation from Jeremiah is strategically devastating. Even the scriptures of the old covenant speak of the necessity of a new covenant that is predicated on a new relationship to the law. With the new covenant, one instituted via the priestly work of the eternal Son, the law that was to guide the people's covenantal fidelity would be inscribed by God into their minds and hearts. In other words, the law itself would be the internal motivating principle for the lives of the faithful, enabling an intimacy and immediacy in their relationship with God that would issue forth in obedience and a deep sense of forgiveness of sin. What is interesting to note here, again, is that though the law's inscription in the minds and hearts of the people stands out, the citation of Jeremiah here seems to function for the author of Hebrews simply as old covenantal affirmation that the old covenant would be nullified and that a "new" one would be instituted that is qualitatively superior to the old one.[3] Hebrews 8:13 stands as a summary statement that clarifies the purpose of the quotation from Jeremiah. There would be a new covenant, and for the author of Hebrews, the new covenant is established through the new high priesthood of Jesus. But the implications of the law

---

[3] Lane, *Hebrews 1-8*, 210.

written on the minds and hearts of the people will find only scant reference in the following chapters.

Here we see, again, evidence of an anthropocentric bias in the argumentation of Hebrews. A passage as robust as Jer. 31:31-34 is apparently cited primarily for its use of one single word: "new." The fact that this is the only place in the whole Old Testament that uses the word "new" to describe the future coming of a qualitatively different relationship with God[4] is mined for its contribution to the developing argument that in the Son is found a superior covenant to the old version. The substance of the blessings attached to this new covenant are largely overlooked. And in the mention of the internalization of the law may be found the basis of an ecological reading of the passage. Here again we will subvert the intention of the author by exploiting the very sources used to buttress the argument in favor of the author's purposes.

Jeremiah, of course, was the key prophet portending the Babylonian Exile of Judah. In ch. 5 of our study, we noted that legislation concerning the Sabbath and Jubilee years for Israel provided for rest for the land in observation of the fact that Israel was to live as tenant farmers in God's garden (cf. Lev. 25:1-12). Moreover, were Israel to fail to observe the Sabbath years of rest for the land, the people would be punished, including expulsion from the land in order to allow the land to recoup the years of rest denied it (Lev. 26:34-35). The Chronicler, writing from the perspective of the post exilic return to the Holy Land, provides an interpretation of history that draws together Jeremiah and Leviticus 25–26. Second Chronicles 36:20-21 reads:

> He [Nebuchadnezzar] took into exile in Babylon those who had escaped from the sword, and they became servants to him and to his sons until the establishment of the kingdom of Persia, to fulfill the word of the LORD by the mouth of Jeremiah, until the land had made up for its sabbaths. All the days that it lay desolate it kept sabbath, to fulfill seventy years.[5]

Here we see that a contributing factor to the Babylonian Exile was failure to abide by Sabbath year legislation. Interestingly, among both biblical writers and subsequent interpreters, the factor that led to the exile that receives most attention is idolatry. Yet in the Chronicler we see a strand of intra-biblical

---

[4] Lane, *Hebrews 1–8*, 209.
[5] For discussion, see Van Dyke, *Between Heaven and Earth*, 62.

interpretation that connects Jeremiah's prophecy regarding the duration of the exile with Sabbath legislation. If we may indulge in a bit of presumptive arithmetic for a moment, assuming there were literally seventy missed Sabbath years (not counting Jubilee years at this point), then that would mean the people went nearly five centuries without observing Sabbath legislation. This would surely count as evidence for the ineffectual nature of the law as declared by the author of Hebrews.

If we were to follow the lead of the Chronicler, we might also connect Jeremiah's prophecy to Sabbath legislation, but instead of focusing on the seventy-year duration prophecy, we would focus on the prophecy quoted by the author of Hebrews. And we would go beyond the author of Hebrews by assigning the substance of this prophecy to the new law brought about with the change in priesthood. We would also go beyond our author's appropriation of the passage by focusing on the efficacy of the law written on the minds and hearts of the faithful.

It is noteworthy that the actual word "law" is used in the Jeremiah quotation, given that the author of Hebrews seems reticent to use it in the context of the new covenant. As we noted, "oath," "promise," and "hope" were all used in the discussion of the new law necessitated by the new priesthood. Yet in the quotation, the term "law" is front and center, though again relegated to non-mention by the reticence of the author. Here an ecological reading would gladly appropriate this term and mine the law for its contributions on behalf of Earth, which would lead directly to Sabbath legislation. The difference this time, however, is that the law to which we refer is no longer what was written on stone tablets, external to the people, but is rather written directly on the minds and hearts of the faithful (cf. 2 Cor. 3:3). In such a configuration, attention to the land's well-being would not reside in weak statutes, but would be intrinsic to the motivations of the faithful. In the same way that the Chronicler interpreted Sabbath legislation in light of Jeremiah's prophecy from his own perspective in history, so too would we bring Sabbath and Jeremiah together from our own perspective in history, one characterized by ecological crisis.

Hearing Earth

Two main themes occupy Earth's focus in Hebrews 8. First, the idea of the "shadow" arises. As put forth by the author of Hebrews, the earthly sanctuary is but a shadow of the heavenly sanctuary. In the argument, this obviously speaks to its inferiority to the heavenly sanctuary, largely because it was an artifice of human manufacture. It must be said here that though the earthly sanctuary was a human construction, it was nevertheless constructed at the command of God. It was never to be the ultimate expression of divine presence in the world, but it had a purpose in God's economy. And if it was a humanly constructed sanctuary, it was constructed from materials at hand, and those materials were derived from Earth. Rather than allow an implicit denigration to stand as argued in the letter, Earth would have us understand that Earth shared its resources with human beings in order to help them to follow the divine command to build a sanctuary. But we must also make sure to follow the full import of the imagery of the shadow here.[6] Of course, a shadow is not the same as the thing from which the shadow is formed. And a shadow creates a darkness that may not allow full vision of the thing casting the shadow. However, if followed, the shadow will lead to that which casts the shadow. The earthly sanctuary portends to that which is to come, in Hebrews that being the heavenly sanctuary. But Earth would remind us that the final dwelling place of God will be among human beings on the new earth in the New Jerusalem (Rev. 21:1-4), where no temple exists because the temple is "the Lord God the Almighty and the Lamb" (Rev. 21:22). In this respect, both the original earthly sanctuary relativized by the author of Hebrews and the heavenly sanctuary that the author lauds may both function literarily as shadows of the final dwelling place of God, Earth restored to its original destiny.

Another area in which Earth would have us focus is that of the law. For the author of Hebrews, the law is tied inextricably with the old covenant, and as we saw in Hebrews 7, the law, unable to make anything perfect, gave way to a better expression. Yet as we saw, the author left the door open for reconsideration of the law by quoting Jer. 31:31-34. How might Earth direct us at this point? By looking at how Jesus "fulfilled" the law. Of course, the author of Hebrews has given an extended discourse on how Jesus might be said to fulfill the law—by becoming that toward which the whole old covenantal economy, law included, pointed. Still, the author repeatedly supports assertions by quoting and/or

---

[6] I discuss this understanding of "shadow" in Heb. 8:4-6 in my study on Hebrews, *The Letter to the Hebrews*, 74–75.

alluding to the old covenant scriptures. So something of validity certainly must remain in these scriptures.

Earth would remind us of how Jesus once framed his understanding of the law in his ministry. In Matt. 5:17, within the Sermon on the Mount, Jesus makes the famous statement, "Do not think that I have come to abolish the law or the prophets; I have come not to abolish but to fulfill." Jesus goes on to describe the ongoing validity and authority of these scriptures, concluding that his followers must exhibit a greater righteousness in their lives than do the Pharisees (5:20). Then follows a series of "antitheses" where Jesus contrasts his own radical interpretations of certain commandments, illustrating how this greater righteousness works itself out. Earth might ask us to consider how we might extend the commandment regarding the Sabbath: "You have heard that it was said, 'Remember the Sabbath day, and keep it holy'" (cf. Exod. 20:8). This might cause us to recall the substance of this command from the Decalogue given in Exod. 20:9-11. Or it might cause us to see how this was developed with respect to the land in Lev. 25:1-12. But more importantly, how would we, in the spirit of Jesus's sermon, follow the words, "But I say to you"? Since what Jesus was doing in this part of the Sermon on the Mount was getting at the true intention of the law, what would be the true intention of the Sabbath commandment? Perhaps Earth might model this for us in this manner: "You have heard that it was said, 'Remember the Sabbath day, and keep it holy,' and that you must give the land its Sabbaths as well, but I say to you, you shall actively help the land to thrive and attain its destiny to be God's dwelling place." Here the focus is active rather than passive. What might we human beings do, not just to cease exhausting the land but actively to nurture and care for the land? Earth reminds us to probe more deeply into the law's intent, but the law is now inscribed in our minds and hearts, as the author of Hebrews has told us, and the power to perform the "just requirement of the law" is enabled by the indwelling presence of the Holy Spirit (cf. Rom. 8:4). We of the new covenant must far surpass the fading glory of the old covenant. We do so by reclaiming the gift of the law and looking deep into its heart for its true intentions regarding Earth.

9

A New Order of Worship (Heb. 9:1–10:18)

Contents

The new covenant priesthood established in Christ requires both an altar for priestly service and something to offer. Given the superior priesthood of Jesus, it stands to reason the author of Hebrews would argue that such an altar and offering would be of superior status compared to their counterparts in the old covenant. The basis of the superiority of the new altar would be that the earthly sanctuary is a mere copy, a shadow, of the true heavenly sanctuary. And in a startling assertion, the author argues that the new sacrifice is none other than Jesus himself. His blood, being that of a pure human being, offered as it is on the heavenly altar, effects the purification of the conscience of worshipers in a way that the blood of bulls and goats offered in the earthly sanctuary could not. Viewed from an ecological perspective, this section presents astounding considerations for an ecological reading of the text. First, attention is given to the true function of the tabernacle. In the Genesis creation narratives, as we have already seen, the focus of the Sabbath is the establishment of creation as the dwelling place of God. The earthly tabernacle is a mere shadow of this realization, but it prepares the way for the true realization of this vision, not just in the incarnate eternal Son Jesus but also in the joining together of heaven and earth as the Son enters the eternal sanctuary to offer his own blood. But second, this blood is more than just a substitute for the blood of animals. It represents the offering of humanity back to God to fulfill the reclaimed Adamic vocation. This, ironically, subverts the author's diminishing of the efficacy of animal sacrifice, for it argues for a proleptic reading of the sacrificial system as that which anticipates the restoration of the orders of creation in the Son, and restores human beings as the priestly co-regents with God in creation.

Structure

9:1-10: The true tabernacle
9:11–10:18: The logic of the blood of Christ

Analysis

The True Tabernacle (9:1-10)

The author resumes the discussion of the true tabernacle, having introduced the topic at the beginning of ch. 8 only to interrupt it with the declaration of the establishment of a new covenant as prophesied by Jeremiah. The section 9:1–10:18 is functionally the conceptual center of the epistle, for here we see converging the notions of priesthood, sanctuary, and sacrifice. It also continues the rhetorical strategy of comparing Jesus the Son favorably with elements of the old covenant. But here the theological argument of the letter reaches a crescendo, identifying precisely why it is that the new covenant instituted in the priestly ministry of the Son achieves what the old covenant could not. The fundamental argument is that the blood of the perfect sacrifice is offered by the perfect eternal high priest in the perfect heavenly sanctuary and so is able to purify the consciences of worshipers to worship God rightly, something the repetitive sacrifices of the blood of animals by imperfect mortal priests in a mere shadow of a sanctuary could never do. Moreover, the body of this discussion occurs between citations of Jeremiah's prophecy in Heb. 8:8-12 (Jer. 31:31-34) and Heb. 10:16-17 (Jer. 31:33-34), effectively distilling the substance of the new covenant predicted by Jeremiah in the intervening discussion.

Hebrews 9:1-5 provides a brief description of the constructed space of the earthly tabernacle. The structure consists of two parts. The first described is the Holy Place, in which was found the lampstand, table, and bread of Presence. It was here that the daily sacrifices and rituals of the priests were performed. The second, the Holy of Holies, was the place where the high priest entered once a year to offer blood for his own sins and for the sins of the people. In this layout of the tabernacle, the author of Hebrews sees a "parable" of the ages (vv. 6-10). The Holy Place symbolizes the age when priests go daily to offer ineffectual sacrifices on behalf of the people until such time that the true high priest, the one in the order of Melchizedek, would come to offer once for all the perfect sacrifice that would perfect the consciences of worshipers.

From an ecological perspective, we may first reiterate what we said in the previous chapter of our discussion. While the earthly sanctuary is but a shadow of the heavenly sanctuary, nevertheless God did deign to condescend to human frailty to dwell with the people in a sanctuary constructed with human hands, thus hallowing the elements of creation used to build the tabernacle. We may also reiterate that both the earthly sanctuary and the heavenly sanctuary function to anticipate the time when God would bring to realization the intended design for creation, which was to be the dwelling place of God. But there is a further affirmation we may observe in the present passage.

The author makes note of the twofold construction of the earthly tabernacle as an anticipation of the new covenantal age. In the time of the earthly tabernacle, there was a section in which the priests conducted their daily rituals and offered their daily sacrifices. But, as the author of Hebrews notes, there was the special high priestly offering made annually on the Day of Atonement. All of this, again, anticipates the time when things will be set right (v. 10). Perhaps an ecological reading of this passage might offer another "parable" in the earthly tabernacle. In this new parable, the anticipated Melchizedekian high priesthood of Jesus, who reclaims in this priesthood the priestly vocation of Adam, has gathered around him cadres of priests who perform their priestly duties on a daily basis. If Jesus has restored the Adamic vocation in the salvation secured for human beings, then the priests assembled around him would perform daily the benevolent priestly co-regency with God entailed in the Adamic vocation. They would "serve and protect" (cf. Gen. 2:15) Earth, extending into all creation the blessings of God's Sabbath rest as they help creation to attain to its destiny to be the sanctuary of God. The author's parable is capable of another expression when read ecologically.

The Logic of the Blood of Christ (9:11–10:18)

This section of the epistle is rather dense, covering a great deal of ground in substantial detail. This makes sense since the heart of the whole argument for our author culminates here. The argument has been skillfully crafted to reach this point, where the complete superiority of the new covenant established by the Son is demonstrated powerfully in imagery that is at the heart of the old covenant. The argument itself is rather straightforward. The animal sacrifices of the old covenant were not sufficient to procure salvation for human beings because, at root, they were only sketches (9:23), copies (9:24), and shadows (10:1) of what the Son would accomplish in his high priestly ministry. Given this reality, the author proceeds to make a rather startling assertion. Whereas the prophets of old would attribute the inefficacy of animal sacrifice to human unfaithfulness (e.g., Isa. 1:10-13; Jer. 7:21-24; Hos. 6:6; Amos 5:21-26; Mic. 6:6-8; cf. 1 Sam. 15:22), the author of Hebrews attributes responsibility with the animals themselves.[1] This is an obvious bias against Earth that assumes both christological and anthropological expression. The Son is favored over against the animals and the benefits accrued for human beings take ascendency over against the welfare of the animals. In short, the blood of animals could not remove the effects of sin in human beings, as is evidenced by the need for continual sacrifice, whereas the once-for-all sacrifice of Jesus secures a cleansing of sin such that the human worshipers will have a clear conscience so that they may become true servants of God. So how does an ecological reading proceed at this point?

In my earlier volume, I titled a chapter on animal sacrifice, "What's with Cutting up All Those Animals? Reading the Sacrifice of Christ in Hebrews from the Perspective of the Animals."[2] The point of departure for the chapter was the query as to why the almighty Creator of the universe was so concerned with meticulous details for dismembering and offering animals in sacrifice on behalf of human beings. The chapter began with a broad overview of the biblical depiction of animals, concluding that the biblical depiction was quite ambivalent. On the one hand, in the creation narratives of Genesis 1 and 2, God is shown as creating the animal kingdom "good" and lovingly creating animals to serve alongside human beings. However, there are those passages where the animals are shown as experiencing judgment as a result of human sin in the flood narrative (Gen. 6:7) and the animals will be wiped off the face of the earth along with sinful human beings (Zeph. 1:2-3). Even in the New Testament, Jesus makes the statement that compares sparrows with human beings in a way that affirms the superior value of human beings (Matt. 6:26),

---

[1] I discuss this in *The Greening of Hebrews?*, 24–28.
[2] Lamp, *The Greening of Hebrews?*, 21–36.

and the apostle Paul argues that the command in the Torah that prohibits muzzling oxen when they tread the grain was not, in fact, written for the benefit of the oxen, but for human beings (1 Cor. 9:9-10). At first glance, the sacrificial system of the old covenant would qualify as another case study for a less than positive assessment of animals in the Bible. One cannot make a blanket statement that animals are viewed only positively in the Bible.

As the chapter proceeded, it argued that there was a point of identification between human beings due to their common origin from Earth (Gen. 2:7, 19) and the fact that the "breath of life" (נשמת חיים) breathed into the human being in Gen. 2:7 is approximated with respect to the animals in the flood narrative (Gen. 6:17; 7:15, רוּח חיים; Gen. 7:22, נשמת רוּח חיים). In terms of the argument I have been making in the present volume, if in his incarnation the Son has taken upon himself human nature, he has also taken upon himself the stuff of Earth from which human beings derive, thus identifying not only with human beings but also with all creatures similarly derived. The animals would also share in the redemption awaiting human beings. Indeed, in a very literalistic way, the sacrifice of the Son did redeem the animals with its discontinuation of animal sacrifice.

For this discussion, however, I wish to argue in a different, complementary direction. Several lines of investigation will occupy us here. First, we will look at the affirmation that the Son was given a body to do the will of God (Heb. 10:1-10). Second, we will examine Jesus's entrance into the heavenly sanctuary to offer his blood to God (9:11-12, 23-24). Third, we will offer a subversive reading of the references where the author of Hebrews apparently denigrates the efficacy of the blood of sacrificial animals (9:12, 13, 19; 10:4). If this section of Hebrews serves as the climax of the argument for our author, it will also serve for our ecological reading as a place where the major threads of our reading converge to reinforce the contours of the argument we have presented to this point.

In Heb. 10:1-10 the author states the basic premise of the section. The blood of animals is insufficient to cleanse human beings of sin. The primary proof of this assertion is the continual repetition of the sacrifices. If they were effective, they would not need to be repeated. The central feature of the argument is a citation from Ps. 39:6-8 (LXX; Hebrew 40:7-9). The LXX has a significant alteration from the Hebrew text in the second line of the psalm quoted in Heb. 10:5. In the Hebrew, the line reads, "You have dug an ear for me" (Ps. 40:7), whereas in the LXX the text, as quoted in Heb. 10:5, reads, "You have made a body for me" (Ps. 39:6). Though "ear" would have fit well with the author's focus on obedience throughout the epistle, the choice of "body" here accords well with the emphasis on the Son's total obedience in the entirety of his bodily existence.[3] This takes on significance given how the author has introduced the quotation. In Heb. 10:5, the author does two things. First, the quotation is placed on the lips of Jesus. So when, in v. 7, the speaker of the psalm says, "Then I said," as used by our author, it is Jesus who is speaking. The second thing the author does is place the saying at the time when the Son came into the world. This frames the language of the psalm in incarnational terms.[4] What is taking place in the quotation is that the Son, upon coming into the world, is announcing that in the totality of his earthly existence, he has come to do the will of God. For the author of Hebrews, this demonstrates that the Son would not only be a perfect high priest, but that in every aspect of his human life, he would be obedient to the will of God, and this qualifies him to be the perfect sacrifice. The blood of animals pales to this sacrifice on two counts. First, animals are not human beings. We will have more to say about this later. But here we may say that animals cannot accomplish what is necessary for what is a quintessentially human problem: sin. Second, the problem with animal sacrifices is they cannot cleanse the conscience of human worshipers. But in Jesus, here is a human being who is himself pure, and may thus offer himself to purify human beings. Each of these points figures prominently into an ecological reading of the passage.

We have repeatedly highlighted the importance of the incarnation for an ecological reading of Hebrews. As noted in the opening line of the quoted psalm (Heb. 10:5), God does not desire sacrifice and offering, and as reiterated in the line quoted in v. 6, God does not take pleasure in whole burnt offerings and sin offerings. Rather, what God desires is a human being, a genuine human being, who will do the will of God. The Son, via his incarnation, satisfies this desire of God. He becomes what Adam was to have been, a human being who carries out his divinely sanctioned role in the world. As we have stated on numerous occasions, the vocational mandate for Adam was to serve as God's priestly co-regent in creation to spread God's benevolence throughout the

---

[3] Johnson, *Hebrews*, 251–52.
[4] William L. Lane, *Hebrews 9–13*, WBC (Dallas: Word, 1991), 262.

world. However, Adam failed in this vocation. What the Son does is to replay, in effect, the Adamic project so as to put the project back on track. The Son becomes the Second Adam by coming into the world in order to embody what the original Adam was to be. He does everything that the first Adam failed to do, namely, to live as a genuine human being in obedience to all that God has tasked human beings with doing.

As pertains to the animals, by living in obedience to God's will, Jesus eliminates the necessity for animals to be sacrificed for human sin. As our author states in this section of the letter, the blood of animals is ineffective to deal with the issue of sin. Jesus has, in his incarnation, mitigated the necessity for sacrificing animals. For our ecological reading, at this point, Jesus has restructured the relationship between human beings and animals. Animals are no longer needed to have their lives taken to deal with sinful human beings. The template for genuine human living has been demonstrated in the life of Jesus. And his own sacrifice provides for all human beings to live according to this template, to do what God truly desires, which is obedience and worship empowered from a pure conscience. Jesus's incarnation, seen from this perspective, has established the framework for human existence, which in turn reconfigures the relationship between human beings and animals.

Having lived as a perfect human being and become the perfect, eternal high priest according to the order of Melchizedek, it was necessary that he execute this priesthood by offering a perfect sacrifice in a perfect sanctuary. Hebrews 9:11–10:18 begins with a statement to this effect. Jesus made entrance into the "greater and perfect" tent, which again is characterized as "not made with human hands" (οὐ χειροποιήτου), to which is added the explanatory comment, "that is, not of this creation" (τοῦτ' ἔστιν οὐ ταύτης τῆς κτίσεως), in order to offer his blood to obtain eternal redemption (9:11-12). In vv. 23-24, following an interlude in which the author explains that blood was necessary to purify all things required for the worship of God (vv. 18-22), the author asserts that the earthly tabernacle, as a mere "sketch" or "copy" of the true sanctuary, required blood for purification, but the heavenly sanctuary required better sacrifices. Again, the heavenly sanctuary is not one made with human hands (χειροποίητα), but is actually heaven itself.

We should not think that the blood of Jesus is used to purify the heavenly sanctuary, as the blood of animals was for the earthly sanctuary. Rather, the focus is here on what is offered in the heavenly sanctuary. It is the blood of Jesus himself. For the author of Hebrews, this perfect offering in the perfect sanctuary by the perfect priest is able to secure eternal redemption for human beings, precisely because it is the blood of a human being. But how should we think about this ecologically?

Even though the passage itself began with the assertion about the blood of Jesus being offered in the heavenly sanctuary, logically prior is the incarnation of the Son in order to do the will of God and become the perfect offering for sin. That is why we discussed the incarnation first. For had the Son not lived in this way, his blood would not have achieved what it did. For an ecological reading of this notion, we need to focus on the fact that this is the blood of a human being that is offered to God. Another way of looking at this offering of Jesus's blood to God, beyond the eternal redemption touted by the author of Hebrews, is to see that in this offering Jesus is effectively offering human beings back to God. Following what Ben Witherington has called the "blood principle of Lev 17:11," we understand that in offering his blood to God, he is offering the life that is in the blood back to God.[5] In other words, Jesus, through his own perfect blood, is offering back to God human life that is lived in obedience to God. As the Second Adam who fulfilled the destiny of human beings to live obediently and perfectly before God, Jesus is offering all human beings back to God to fulfill the destiny of their creation and calling.

Moreover, the framing of this offering is suggestive. In Heb. 9:14, it is through the "eternal Spirit," in all likelihood a reference to the Holy Spirit,[6] that the Son offers his own blood to God. Here we see language that would eventuate in the dogma of the Holy Trinity. In this depiction, human blood, Jesus's own human blood, in which resides the principle of human life, is taken up into the life of the Trinity. Since this offering procures eternal redemption for human beings, now human beings receive the purified conscience that enables them to receive the law written on their minds and hearts (cf. 8:10) and to live according to the law. In their lives, human beings may now live as did Jesus, fulfilling the Adamic vocation, and through their lives in the world on behalf of the world, work to bring to the world the blessings of life taken up into the life of the Triune God. As those who in their creaturehood are identified with Jesus, who himself assumed the stuff of Earth in his incarnation, human beings and other-than-human creatures are in principle, in the blood

---

[5] Witherington III, *Letters and Homilies for Jewish Christians*, 273.
[6] See my discussion in *The Greening of Hebrews?*, 59–63, for discussion of this identification and its implications.

of Jesus, offered back to God to live in accordance with the orders of existence established by God in the original design for creation.

Finally, let us turn our attention to the occurrences in 9:11–10:18 of the mention of the blood of animals. In 9:12, the author notes that Jesus did not enter the heavenly sanctuary with the "blood of goats and calves." In the following verse, the "blood of goats and bulls" and the "sprinkling of the ashes of a heifer" are mentioned as things that merely sanctify the flesh of those offering them. In 9:19, Moses is said to have taken the "blood of calves and goats" and sprinkled the scroll and the people for ritual purification. Then in 10:4, the author boldly asserts that the "blood of bulls and goats" cannot take away sins. These references are drawn from the backdrop of the Day of Atonement rituals and are set against the offering of Jesus's own blood, proving the superior efficacy of the Son's sacrifice. Of course, this strategy of comparison has been employed from the opening words of the epistle and is used to exhort the audience to hold fast to their profession of faith. If the new covenant inaugurated by Jesus is superior in every way to the old covenant, then it merits faithful adherence. In light of the author's rhetorical strategy, such a comparison makes perfect sense.

However, some commentators see here a rather tendentious appropriation of animal sacrifice by the author. Luke Timothy Johnson argues that the author of Hebrews is being unfair here in the critique of animal sacrifices. As framed in the Torah, the animal sacrifices served only to make human beings ritually pure for purposes of worship. The author of Hebrews has reframed the issue to deal with the problem of sin, which the blood of animals could never address, for that was never the intention behind the sacrificial apparatus of the old covenant. Given that our author has so reframed the issue, something other than the blood of animals is needed.[7] Harold Attridge is rather blunt in his assessment of the author's critique of the animal sacrifices, calling it "deprecatory generalizing."[8] So what is the author of Hebrews not telling us about the animal sacrifices?

Some current thought is directed toward the strong underlying assumption of the sacrificial system of the old covenant.[9] In a thought provoking essay, Jonathan Morgan assesses the sacrificial system as described in Leviticus

---

[7] Johnson, *Hebrews*, 249.
[8] Attridge, *The Epistle to the Hebrews*, 248.
[9] I address this in *The Greening of Hebrews?*, 28–30.

from an ecological point of view.[10] Morgan notes that the prescriptions that the animals sacrificed must be without spot or blemish highlight the fact that human beings are not without spot or blemish, and thus are of themselves not able to bridge the barrier between the earthly and heavenly realms. To bridge this gap is a deadly transaction, and it is unthinkable that a human being would be sacrificed for this purpose, not only because such is outside the sacrificial protocols of ancient Israel but also because the human being is not of proper purity to do so. In this respect, the sacrifice of a spotless and unblemished animal bridges the gap and allows right ordering of the social order within the covenant community. This understanding shifts the focus away from understanding the animals as "victims" offered in sacrifice to absorb the punishment that is due human beings for their transgressions. Rather, the animals are seen as holy things performing a holy service that human beings are simply unable to perform for themselves. This understanding highlights the gift that animals bestow on human beings. So in one sense, the author of Hebrews is correct; animal sacrifice cannot ultimately provide a solution for the sin that plagues human beings. Thus the necessity for the offering of a perfect human being to restore human beings to their rightful place in the created order, as we have described earlier. Michael Northcott observes here a proper valuation of the role of animals in restoring the relationship between the human and divine.[11]

However, another factor comes into play here. As we have sketched it, the proper role of human beings in God's created order is for human beings to function as the priestly co-regents with God on behalf of creation. As it stands, the sacrificial system actually depicts an inversion of this creational ideal. In the sacrifice of animals, it is the animals that are performing this mediatorial role on behalf of human beings. The sacrificial system stands as a stark witness to the disruption of the God/human/other-than-human triad as originally established. With the offering of Jesus's own blood, in which Jesus effectively offers to God that which represents the life of all human beings, the order

---

[10] Jonathan Morgan, "Sacrifice in Leviticus: Eco-friendly Ritual or Unholy Waste?" in *Ecological Hermeneutics: Biblical, Historical, and Theological Perspectives*, eds. David G. Horrell, Cherryl Hunt, Christopher Southgate, and Francesca Stavrakopoulou (London and New York: T & T Clark, 2010), 37–43.

[11] Michael S. Northcott, *The Environment and Christian Ethics*, New Studies in Christian Ethics (Cambridge: Cambridge University Press, 1996), 186.

within the triad is reestablished. Jesus fulfills that which is foreshadowed in the sacrifice of animals. In doing so, Jesus restores the Adamic priestly vocation to human beings and frees animals from bearing a burden they were never meant to bear. Read from this ecological perspective, animal sacrifices should not be deprecated in the manner suggested by the author of Hebrews. Rather, they should be celebrated for their mediatorial service on behalf of human beings, for the repair of the divine-human relationship they offered until such time as the Son would come to restore human beings to their rightful place as those who would protect and serve the whole created order in priestly service to God.

Hearing Earth

One way in which we might hear the voice of Earth regarding the animals is in a call to reassess our ethical approach to animals. Many Christian thinkers are currently engaged in the task to describe an approach to animals that reflects the value of animals in the order of God's creation. At a biblical level, attention might be drawn to God's assessment of the creation of other-than-human living beings as good in the opening chapters of Genesis. Or we might read Psalm 104 or Job 38–41 where God is shown to delight in the animals. But as we noted earlier, not all biblical data assert an unqualified positive view of animals. The Christian tradition, though also ambivalent in its assessment of the treatment of animals, has resources from which to draw in crafting an animal ethic. St. Francis of Assisi is known as a patron saint of animals. John Calvin,[12] John Wesley,[13] and C. S. Lewis[14] have all argued for the dignity of animals and urged their compassionate treatment. More recently, attention is focused on the theological bases for ethical treatment of animals. Laura Hobgood-Oester argues for an ethic that seeks to reclaim from Christian tradition a view of animals such that the gospel may be seen as good news for all creatures.[15] Michael Hogue argues for an approach that seeks to delineate

---

[12] Peter A. Huff, "Calvin and the Beasts: Animals in John Calvin's Theological Discourse," *Journal of the Evangelical Theological Society* 42 (March 1999): 67–75.

[13] Michael Lodahl, *God of Nature and of Grace: Reading the World in a Wesleyan Way* (Nashville: Kingswood, 2003), 195–202.

[14] Andrew Linzey, "C. S. Lewis' Theology of Animals," *Anglican Theological Review* 80 (Winter 1998): 60–81.

[15] Laura Hobgood-Oster, *The Friends We Keep: Unleashing Christianity's Compassion for Animals* (Waco: Baylor University Press, 2010). A similar approach is found in Andrew Linzey, *Animal*

the relationship between human beings and animals in a complex web of interactions that takes account of the unique position of human beings while acknowledging the biological connections between human beings and the animal world.[16] Richard Wade, arguing from a natural law perspective, seeks to balance the "duty to do what is ethically right with respect to the nature of the animal (the kind of being the animal is)" while at the same time assenting to "duties to human persons with a reasonable need."[17] This sampling of the literature demonstrates that concern for animals is both rooted in the sources of Christian tradition and is a thriving current interest among ecologically concerned Christians. Earth would exhort us to pursue this with due diligence.

In this light, Earth may direct our attention to one specific manifestation of societal disregard for animals that figures prominently into the lifestyles of many Christians: the industrial meat production system. Over the course of the past half century or so, there has been a shift in meat production from many family farms to a relatively small number of industrial factory farms operated by large, multinational agribusinesses. These factory farms produce large numbers of animals constrained in what are called concentrated animal feeding operations, or CAFOs. In such a business model, animals are nothing more than a resource to be exploited to meet the growing demand of industrial societies for inexpensive meat. While predominantly an issue for the economies of the Global North, with the rising economic power of highly populous nations such as China and India, the desire to share in the standards of living of those in the Global North is creating huge demands for access to meat. The living conditions of these animals was exposed in the documentary *Food, Inc.* (2008) and in a volume titled, *CAFO: The Tragedy of Industrial Animal Factories*.[18] In this collection of essays, authors explore the effects of CAFOs on animal health and welfare, water and land quality issues, environmental justice implications, global climate change, and economic exploitation of human workers. In light of a system that relegates animals to the status of a consumer commodity, Earth cries out in protest against the abuse of this segment of

---

Gospel: Christian Faith as if Animals Mattered (Louisville: Westminster John Knox, 2000). See also Bauckham, *Living with Other Creatures*.

[16] Michael S. Hogue, *The Tangled Bank: Toward an Ecotheological Ethics of Responsible Participation* (Eugene: Pickwick, 2008), 14–21, 235, 244.

[17] Richard Wade, "Towards a Christian Ethics of Animals," *Pacifica* 13 (June 2000): 202–12.

[18] Daniel Imhoff, ed., *CAFO: The Tragedy of Industrial Animal Factories* (Los Angeles: Foundation for Deep Ecology, 2010).

the Earth community and demands justice. Earth calls on the followers of Jesus to wrestle seriously with the ecological implications of his sacrifice and the ramifications of it for both animals and human beings. In regard to the animals, Earth calls for human beings to acknowledge the debt due to those creatures who were sacrificed in times past to facilitate covenant relationship with God and to accept their status as the co-redeemed by Jesus's sacrifice. As for human beings, Earth would call out to them to accept their reclaimed status as the priestly intermediaries between God and the other-than-human creation and to protect and serve the animal constituency. This would entail the discipline to curtail gluttonous tendencies to consume animals beyond what is needed for human survival and to find ways to participate with animals in the biosphere to ensure thriving for all of God's creatures.

10

Looking for the City Whose Foundations Are in Heaven (Heb. 10:19–11:40)

Contents

The theological substance of the author's argument for the superiority of the Son, and thus the new covenant, over the old covenant now finished, the author moves to the exhortation of the people to demonstrate the faithfulness of the patriarchs who excelled in faith under an inferior covenant. Following another dire warning not to forsake the new covenant in Christ, the author runs down a sort of "roll call of faith," drawing on some expected as well as highly surprising figures to show that faithfulness, possible under the old covenant, should be much more expected under the new. Ecologically, ch. 11 does two things. First, it establishes a theological reading of the covenantal roles that several of the named figures play in the reestablishment of the Adamic vocation. Second, it lays important groundwork for the final stage of the letter, where the goal of creation to become the dwelling place of God is described.

Structure

10:19-39: Yet another warning to hold fast to the new covenant
11:1-40: The covenantal progress toward the new city

Analysis

Yet Another Warning to Hold Fast to the New Covenant (10:19-39)

With Heb. 10:19, the author moves from the rhetorical strategy of explicit comparison of the new covenant instituted by the Son with aspects of the old covenant to an exhortation for the audience to remain faithful to the new covenant, knowing that perseverance brings with it eternal reward. With "therefore" (οὖν), the author begins a lengthy discussion of the ramifications of all that has gone before, first focusing heavily on exhortation (vv. 19-25) and then shifting to a dire warning (vv. 26-39). The net effect of this section of the epistle is that the people must be faithful to the one who has called them into this covenant, even in the face of persecution, for to become unfaithful would be to suffer the rejection of those who have come before who were disobedient and unfaithful (cf. 3:7-4:11).

Within this exhortation and warning there are a couple of points of interest that contribute to an ecological reading of the epistle. Having just spoken at length about the blood of Jesus (9:11-10:18), the author identifies as a potential offense of the people "profan[ing] the blood of the covenant by which they were sanctified" (10:29). The translation of the NRSV "profaned" renders the Greek κοινὸν ἡγησάμενος, which literally means "reckon common." Given the profound significance of the Son's blood in the economy of salvation, such a reckoning of Jesus's blood would certainly count as blasphemous. In our ecological reading of the blood of Jesus, namely that it represents an offering of human life back to God so that human beings would reassume the mantle of the Adamic vocation, such an attitude toward the blood of Jesus effectively repeats the sin of the first human beings when they forsook their vocation. Membership in the new covenant, on our ecological reading, entails enlistment in the order of Adamic priestly co-regents with God toward creation. To "reckon common" that which reconstitutes the human vocation toward creation may not ascend to the severity of the offense described here by the author of Hebrews, yet it nevertheless constitutes a rejection of something that is implicit in the effects of Jesus's blood for those whose trust in its efficacy procures salvation. One may not simply accept only those things that seem appealing in the covenant; one must also accept the responsibilities that go along with it.

Perhaps a more cogent point of ecological interest is found in 10:32-39. Here the author encourages the audience to hold fast to their first confession, recalling those times when they suffered various forms of persecution for their faith. In today's world, many who would stand against environmental degradation and ecological destruction face persecution for their prophetic calls to accountability. Just a few days before I wrote this paragraph, a news article came across my desk documenting this trend. The Yale University School of Forestry and Environmental Studies publishes an online magazine, *YaleEnvironment360*, that highlights environmental issues. On July 30, 2019, the magazine reported on a study by the environmental watchdog group Global Witness in which it reported that 164 activists were killed in 2018 as a result of their defense of land and waterways.[1] Of those 164 deaths, eighty-three occurred in Latin America, sixty-four in Asia, fourteen in Africa, and three in Europe (Ukraine), with mining and agribusiness sectors accounting for forty-three and twenty-one deaths, respectively. The Philippines, with thirty deaths, Columbia, with twenty-four deaths, and India, with twenty-three deaths, were the top three most deadly countries for environmental activists. Another study indicates that from 2002 to 2017, the rate of people killed trying to defend water and land resources and wildlife rose from an average of two to four per week, with a total of 1,558 people killed.[2] Often these activists face off in classic David-versus-Goliath scenarios against powerful corporate entities that enjoy support from governmental officials. It is noteworthy that none of the deaths reported in the *YaleEnvironment360* article occurred in the countries of the Global North. As a result, the urgency of the issue is often not noted in those countries. In the United States, for example, activism is often marked in recycling drives and individual lifestyle changes, though significant protests that entailed risks to personal safety, such as the Dakota Access Pipeline protests in 2016, have occurred. Reclaiming the Adamic vocation may indeed require those who act on behalf of Earth to risk life and reputation

---

[1] "164 Activists Were Killed Defending Land and Water Last Year," *YaleEnvironment360*, July 30, 2019. Available online: https://e360.yale.edu/digest/164-activists-were-killed-defending-land-and-water-last-year (accessed August 3, 2019). For specific cases, see "Environmentalists at Risk: An E360 Series," *YaleEnvironment360*, February 27, 2017, March 7, 2017, March 13, 2017. Available online: https://e360.yale.edu/series/environmentalists-at-risk (accessed August 3, 2019).

[2] Nathalie Butt, Frances Lambrick, Mary Menton, and Anna Renwick, "The Supply Chain of Violence," *Nature Sustainability* 2 (2019): 742–47.

in fulfillment of this vocation. Environmental martyrdom may indeed become the reality for those so seeking to live the reclaimed Adamic vocation.

The Covenantal Progress toward the New City (11:1-40)

Hebrews 11 is frequently characterized as the "roll call of faith" or the "faith hall of fame" or some such homiletically motivated description. The author of Hebrews will refer to those named in the chapter as a "cloud of witnesses" in 12:1. However one may choose to refer to the characters of ch. 11, it is clear that they are to stand as examples of faithfulness for the readers to emulate and even surpass, for the audience of this epistle has access to better promises and a better hope through the new covenant instituted by Jesus. In a chronologically ordered presentation, the author of Hebrews takes us from the creation of the world through significant characters of primeval history to major figures in the history of the covenant people, illustrating along the way how they demonstrated faithfulness through difficult circumstances.

The motivation ascribed to these heroes of faith is one that frequently militates against ecological responsibility on the part of people of faith. In the key passage of our concern, Heb. 11:8-16, Abraham, the exemplar of old covenant faithfulness, is said to be looking not for an actual, physical homeland but rather for a "better" and "heavenly" city, prepared for them by God (v. 16; cf. v. 10). Such a characterization of the Christian hope has been appropriated and expanded into an escapist eschatology in which adherents look to the future return of Jesus, frequently cast in terms of a "rapture" from the world that is destined for destruction in order to dwell in disembodied bliss in heaven. N. T. Wright subjects this type of eschatology to scathing critique,[3] attributing its predominant contours to Hellenistic and sometimes Gnostic ideas overlaid on biblical terminology.[4] The implications of an escapist eschatology for ecological responsibility in the present are often disastrous. Alan Cadwallader has characterized such views that render Earth obsolete as "transmitting disastrous ecological consequences and ethical immunities on their perpetrators."[5] Put more colloquially, why would one paint a car that is soon to be driven off a cliff?

---

[3] Wright, *Surprised by Hope*, 118–34.
[4] Wright, *Surprised by Hope*, 88–90.
[5] Alan H. Cadwallader, "Earth as Host or Stranger? Reading Hebrews 11 from Diasporan Experience," in *The Earth Story in the New Testament*, eds. Norman C. Habel and Vicky Balabanski (London: Sheffield Academic, 2002), 149.

In my earlier volume, I examined the crucial passage for an ecological reading, 11:8-16, in the context of a significant trajectory within the letter that has contributed to a biased reading against Earth within the context of escapist eschatology.[6] Following Kenneth Schenck, we might call this trajectory "the transience of the created order."[7] Three passages constitute this trajectory: 1:10-12; 11:8-16; and 12:18-29. We have already addressed 1:10-12, which is a quotation of Ps. 102 [101 LXX]:26-28, in an earlier chapter, and we will address 12:18-29 in detail in our next chapter. The point of each of these passages, however, is that the created order is depicted, in some way, as transient or temporary, at least in comparison to the eternity of the Son or God or the heavenly realities. How is this depicted in 11:8-16?

The key word in ch. 11 is "faith" (πίστις), used twenty-four times in the chapter. Within this number of occurrences, the trope used throughout ch. 11, "by faith" (πίστει), accounts for eighteen of these occurrences. The term is typically fronted before the name of a person who exemplified faith by some action. In light of the working definition provided by the author of Hebrews in v. 1, where faith is described as the "assurance of things hoped for, the conviction of things not seen," those who operate on the basis of faith set their eyes on things that are not visible and base their actions on attaining those unseen things. Indeed, "by faith" we know that the worlds themselves were created out of things unseen (v. 3). So when we arrive at v. 8, where it is said that Abraham set out from his homeland for a place he was to receive as an inheritance, he did so by this faith that takes into account the unseen and acts toward its attainment. Indeed, a subtle vocabulary change occurs in the author's appropriation of the calling of Abraham in Gen. 12:1. In the LXX, Gen. 12:1 notes that then Abram was told to go to a "land" (γῆ), which clearly denotes a physical geography, whereas in Hebrews, the author has shifted the term to "place" (τόπος), which may indicate a "transcendent site, esp[ecially] the place to which one's final destiny brings one."[8] Given that Abraham is said to have lived in that ostensibly promised physical geography as a foreigner (Heb. 11:9) as he looked for a "city that has foundations, whose builder and architect is God" (v. 10), it is clear that the author has substituted

---

[6] Lamp, *The Greening of Hebrews?*, 72–76.
[7] Schenck, *Cosmology and Eschatology in Hebrews*, 122–32.
[8] BDAG, 1011.

a transcendent destination for a terrestrial one.⁹ And in vv. 14-16, the author explicitly contrasts a terrestrial destination with a heavenly one, noting that if Abraham had indeed been looking for an earthly destination, he could have easily returned to his own homeland, but instead, he was seeking a heavenly country.

So it appears that the author of Hebrews has effectively spiritualized the promise of a land, as with the promise of the land given to Joshua (4:1-11).¹⁰ At this point, it may be tempting to seek a subversive reading that counters the assertion that the author of Hebrews has established an eschatological paradigm that renders the physical creation obsolete and destined for annihilation. We will offer such a reading, but it will have to wait for our discussion of the final passage in the trajectory of "transience of the created order," 12:18-29, which will occur in the next chapter of our study. In the meantime, we will settle for an approach to 11:8-16 as it stands that seeks an Earth-friendly appropriation of the passage.

Here we turn to the characterization of Abraham as living "in a foreign land" (v. 9). An array of vocabulary is used to depict Abraham, and his descendants after him, as "resident aliens" with respect to this world.¹¹ In v. 9, the verb παροικέω conveys the sense that Abraham migrated to a place where he lived as a stranger.¹² In v. 13 they confessed that they were "strangers" (ξένοι)¹³ and "foreigners" (παρεπίδημοι)¹⁴ "on the earth" (ἐπὶ τῆς γῆς). Note here that the author uses the term γῆ ("earth") to locate these patriarchs in a terrestrial locale, in contrast to the transcendent destiny toward which they proceed. This vocabulary firmly establishes that Abraham, and in the argument of the author of Hebrews, the people of the new covenant, live as resident aliens in a land that is ultimately not their homeland. As Norman Habel notes, this image is rendered in the Abraham narratives of the Hebrew Bible as an "immigrant ideology" regarding the land in which the land serves as a host country where Abraham would dwell as he looked forward to the future homeland of his descendants and in which there would be peaceful

---

⁹ So Lane, *Hebrews 9–13*, 349; cf. Craddock, *Hebrews*, 136.
¹⁰ Compare Lamp, *The Greening of Hebrews?*, 71–72. See also ch. 5 of this current study.
¹¹ Koester, *Hebrews*, 485.
¹² BDAG, 779. Note that in 1 Pet. 1:1 this verb is used and in 1 Pet. 2:11 the cognate noun πάροικος occurs to connote the people of God as pilgrims on earth.
¹³ BDAG, 684.
¹⁴ BDAG, 775.

relations with the inhabitants of the land.[15] As inhabitants of another's land, the resident aliens would be interested in the well-being of that land, if for no other reason so that their habitation would be more pleasant as they anticipate their future homeland. They would "strive for the gradual though never final approximation of its qualities in whatever locale one happens to live."[16] Even as "resident aliens" on earth, the people of God, regardless of their eschatological position, should desire the best possible conditions for life while they inhabit earth, including the ecological well-being of Earth.

Alan Cadwallader offers a sort of flipside to the concept of the people of God as "resident aliens" on earth. He draws attention to the notion of Earth as a "host country." Cadwallader notes a subtle intertextual echo in the author's use of the phrase "on the earth" in v. 13. He observes that in Gen. 24:37, which focuses on Abram's dwelling in Canaan, the language used there, "in their land," has actually been expanded by the author of Hebrews to include the whole world as being the dwelling of the resident alien Abram.[17] In Cadwallader's reading, the terms "heavenly" and "better" in Heb. 11:8-16 describe a qualitatively different encounter with Earth, leading to an ecologically conscious response to Earth that appreciates the hospitality of Earth to its inhabitants. A mutual engagement of the people of God as "resident aliens" and Earth as a "host country" leads to a concern on the part of the people of God for the well-being of their host, which in the present discussion would entail a concern for the ecological well-being of Earth.

Hebrews 11:8-16 has been the focus of our ecological analysis of this chapter so far. Before concluding our discussion of ch. 11, there are two further matters that contribute to an ecological reading of the letter that demand some attention. The first is a sequence of events/persons mentioned in the chapter, and the second is the matter of Abraham's attempted sacrifice of Isaac.

As we noted in ch. 6 of our study, when we first encountered the name of Abraham in Heb. 6:13-14, Abraham is a key figure for an ecological reading of Hebrews because in the person of Abraham, we find not just the person through whom God's redemptive covenant would be initiated but one whose role in this covenant would be the recapitulation of Adam. Recall we said there

---

[15] Habel, *The Land Is Mine*, ch. 7.
[16] Jewett, *Letter to Pilgrims*, 205.
[17] Cadwallader, "Earth as Host," 159–61.

that the covenantal promise made to Abraham paralleled the vocation given to Adam. As summarized by N. T. Wright, the promises made to Abraham echoed the commands given to Adam in Gen. 1:28. Adam was to be fruitful, multiply, and fill the earth. Abraham was promised that he would be the father of a great nation and inhabit a land (12:2-3; 17:2, 6, 8; 22:16-18). This pattern of promise was repeated to Isaac (26:3-4, 24) and to Jacob (35:11-12), as well as in Isaac's blessing of Jacob (28:3-4) and Jacob's blessing of Joseph (48:3-4).[18]

But Wright also goes further in both directions from the time of Abraham and his posterity. Prior to Abraham, we read the story of Noah and the great flood. In this story, the entire biosphere is destroyed, save for a scant remnant through which the animal world would be repopulated, brought on by God's judgment against sinful humanity. How might we respond from an ecological perspective? Habel has seen this as an instance where Earth is clearly violated in its integrity as a subject, and Earth's voice is heard in protest at its treatment.[19] I wish to present another option from which we might derive a more positive ecological assessment of the account.[20] Focal to our study has been the contention that human beings were created to function as the divine image-bearing priestly co-regents of God in creation to mediate God's benevolence throughout creation in preparation for it becoming the dwelling place of God. This understanding frames the relational triad of God/human/other-than-human creation in a very specific way. All of creation, for better or worse, often the latter, is bound up with human beings and their fulfillment of their vocation. Paul gives tacit affirmation of this in Rom. 8:18-25. The flood narrative provides a vivid example of the deep connection between human beings and the rest of creation as well as a sobering illustration of how deeply God honors the integrity of the relationship between human beings and the rest of creation established in the orders of creation. Viewed piecemeal, God seems to be punishing all of creation for the actions of one rather small component of that creation. Frankly, it does not seem to be just. However, given the orders of creation established from the beginning, God may be viewed as simply honoring those orders. Such an approach is seen frequently in God's future dealings with Israel, with Earth suffering for the failure of the covenant people

---

[18] See Wright, *Paul and the Faithfulness of God*, 783–95.
[19] Habel, *The Birth, the Curse and the Greening of Earth*, 83–102.
[20] Elsewhere I have examined Darren Aronofsky's 2014 film, *Noah*, which I argue is actually a paradigm for reading the Bible ecologically. See Lamp, *Reading Green*, 16–25.

to fulfill the Adamic vocation with which they have been tasked (e.g., Hos. 4:3; Zeph.1:3).

Despite the apparently overwhelmingly negative picture we see in the story of Noah, this is not, ecologically speaking, where the story ends. When Noah found dry land following the flood, God made a covenant with Noah that God would never again curse the ground due to human wickedness, nor would God ever again destroy the other-than-human inhabitants of the earth (Gen. 8:21), but rather the normal cycles of seasons and fecundity would prevail for as long as the earth endures (8:22). In 9:1, Noah was given the same command as Adam before him and Abraham after him: "Be fruitful and multiply, and fill the earth."[21]

The culmination of the Noah story is the mandate given to Noah to resume the Adamic vocation (see Gen. 9:1) by multiplying and filling the earth. However, a crucial caveat is given in this resumption of the vocation. Whereas Adam would rule benevolently, here the harsh reality of Noah's mandate is that the animals would now fear and dread their human overlords (v. 2), and now for the first time, animals would be available for human consumption as food (v. 3). At this stage of God's redemptive program for creation, a sort of "reboot" is in place, but the parameters of the relationship have changed to reflect the reality evidenced in the flood narrative. We might view this as a divine concession that the original orders no longer avail. However, a key restriction is given in v. 4: human beings may not consume the blood of animals. Further restrictions on the human consumption of animals would come with the distinction of clean and unclean animals in the Torah. Lawson Stone has argued that these regulations in Leviticus function to curtail the unbridled consumption of animals exhibited by those nations surrounding Israel, providing for the covenant people an object lesson in their relationship with the animal world.[22] We saw in our earlier discussion of animal sacrifice that animals function to facilitate divine-human communion that would be fully restored with Jesus's sacrifice that would in turn restore the God/human/other-than-human triad. Noah's place in the grand biblical narrative is not

---

[21] Wright, *Paul and the Faithfulness of God*, 787.
[22] Lawson Stone, "Worship as Cherishing Yahweh's World," paper presented at the annual meeting of the Institute for Biblical Research, New Orleans, LA, November 21, 2009.

the final word. Rather, it is a first step in God's reconstitution of the Adamic vocation.

Going forward from the time of the patriarchal lineage, we see the people of Israel characterized in terms of their fruitfulness in the land of Egypt and of occupying the land (47:27; Exod. 1:7).[23] Terence Fretheim sees in the depiction of Israel in the land of Egypt the beginnings of the fulfillment of Adam's vocation, to become the people through whom God will bring the restoration of creation and whose opposition by Pharaoh represents the powers of chaos as exhibited in his antilife measures against Israel.[24] What Wright sees in all of these figures is a trajectory in which the Adamic vocation is both resumed in these figures as well as redeemed in them, culminating in Christ, who as the Second Adam fully reclaims the Adamic vocation for human beings.

When we turn our attention again to Hebrews 11, we see a similar trajectory to that of Wright recounted. Of course, for the author of Hebrews, the real focus here is not on the lineage of covenantal development but on how the major figures of the old covenant exhibit the faith that the author wishes for the readers to follow. For now, we are content to note the author's sequential depiction of these figures. Beginning with an oblique reference to Adam in the mention of the creation of the worlds from the unseen (v. 3), we find Noah recalled in v. 7, with Abraham's example occupying much attention beginning immediately in v. 8, with mention of Isaac and Jacob following in v. 9. The patriarchal story is recounted through v. 23, with mentions of the blessings of Jacob and Joseph in v. 21 and v. 22, respectively. Moses and his story begins in v. 23 and extends through v. 29. Though the author of Hebrews and Wright see different things in their historical overviews, the overviews themselves provide the contours of each author's argument. For the author of Hebrews, the issue is faith; for Wright, it is the resumption of God's rescue mission for creation. For us, we see a way to construe this lineage as the procession of God's mission to restore creation, through God's human creatures, such that the vocational mandate given to Adam to serve as God's priestly co-regents in creation to minister God's benevolence to creation achieves its climactic expression in Jesus the Son.

Of course, the examples of Noah and the Exodus present their own difficulties in reading the Bible ecologically. We will have opportunity to

---

[23] Wright, *Paul and the Faithfulness of God*, 791–95.
[24] Terence Fretheim, *Exodus*, Interpretation (Louisville: John Knox, 1991), 106–11.

examine the Exodus account more closely in our next chapter. For now, another interesting passage from an ecological perspective presents itself in Hebrews 11: Abraham's "sacrifice" of Isaac (vv. 17-19). To this we now turn.

This account has the potential to divert our investigation significantly off course. The interpretive and theological issues are numerous and significant. Not the least of these is the author's appeal to the event not to speak of the sacrifice of the Son but rather to be one of the rare mentions of "resurrection" in the letter, and then only as a figurative allusion to Abraham's belief in the concept of resurrection. Our objective in addressing this passage is quite modest. In the passage we see the obverse of the author's assertion that the blood of animals was ineffective to deal with the problem of sin. We also see an affirmation of the author's contention that only the sacrifice of Jesus the Son was sufficient to account for the problem of sin in human beings.

As we noted in our discussion of the logic of the offering of Jesus's blood, we observed that the use of animals without blemish or defect was necessitated because human beings could not present themselves in such manner before God. The innocent animals played a role of honor in their substitution on behalf of human beings. But ultimately, it would be necessary for a sinless, perfect human being to present human blood as an offering to God. This human being, of course, is Jesus, and Jesus's blood solves the dilemma of the ultimate inefficacy of animal blood offered in sacrifice. Human sin may now be forgiven, but not only that, human beings may now be empowered to worship God with a clean conscience and to live in holiness before God. Our ecological reading of this showed that not only did Jesus procure an effective forgiveness of human sin, but he also offered the totality of human life back to God in fulfillment of the original vocation for which human beings were created. Brought back around to the account of Abraham's attempted offering of Isaac, we may see another, complementary reason for which Isaac was spared. Isaac was not himself pure and without defect, and was thus not a suitable subject for sacrificial offering. As such, Isaac could not be that human being who would ultimately restore the Adamic vocation to human beings. The blood of various animals may not be effective in our author's estimation, but for the purpose at hand, neither could Isaac's blood. If Jesus's sacrifice for sin is indeed part of the larger redemptive fabric of God for the whole of creation, Isaac is simply not a suitable candidate. Of course he would be spared. This is not just a soteriological requirement; it is also an ecological requirement.

Hearing Earth

How might Earth speak to us in Heb. 10:19–11:40? Two topics may be on Earth's agenda. First, Earth might caution us not to get overly concerned with a person's eschatological orientation when reading the Bible ecologically and seeking for ways to engage environmental issues. Eschatology, as we have noted, is a crucial factor in ecological discussions. We might break down eschatological positions with respect to the ultimate destiny of Earth into two basic camps. There are those who see in passages such as Rom. 8:18-25 a belief that Earth, along with human beings, will experience eschatological liberation and salvation. We might call this a "restorationist" or "renewalist" position. There will be a continuity between the present order and the eternal order, with the present order renewed in realization of its destiny to become the dwelling place of God. Another position, which we might call a "destructionist" position, views such passages as 2 Pet. 3:3-13 as advancing the prospect of the present order's ultimate destruction, either to be replaced with a completely new, wholly other creation, or perhaps not to exist at all eternally, with only human beings dwelling with God eternally in heaven. There will be no continuity of the present order with the future order, if such a future order is to exist at all. This is a view typically associated with dispensationalists.

Often times, the views of adherents of these positions break down predictably with respect to advocacy of environmental engagement in the present. We might expect those who see a positive future for the present order to encourage present engagement in environmental action in anticipation of Earth's glorious future, while we might expect destructionists to view environmentalists as agents of evil who wish to divert attention away from the real task at hand, which is to seek the salvation of individual human beings so that they might escape the destruction to come. However, the case is not that simple. On the one hand, Rom. 8:18-25, though typically understood as presenting the case that Earth has a glorious eschatological destiny, does not actually explicitly encourage environmental engagement.[25] One might just as easily conclude that only God can solve the problem of environmental degradation, and since

---

[25] Brendan Byrne, "An Ecological Reading of Rom. 8.19–22: Possibilities and Hesitations," in *Ecological Hermeneutics: Biblical, Historical, and Theological Perspectives*, eds. David G. Horrell, Cherryl Hunt, Christopher Southgate, and Francesca Stavrakopoulou (London and New York: T & T Clark, 2010), 83–93; David G. Horrell, Cherryl Hunt, and Christopher Southgate, *Greening Paul: Rereading the Apostle in an Age of Ecological Crisis* (Waco: Baylor University Press, 2010), 121–26.

God has promised to do so, present action is at best superfluous. On the other hand, there are those who hold to destructionist positions who nevertheless argue that present environmental action is necessary for continued human thriving.[26] Given that we do not know the exact timing of Jesus's return to remove human beings from the planet, it may behoove us to care for the environment because we and our descendants will have to survive, and clean water, air, and productive land will be necessary for that.

In light of these varied ecological responses to eschatology, Earth would simply ask us not to become so concerned with arguing the merits of one eschatological system over another as a basis for encouraging environmental action in the present. This diverts from the real issue, which is to get people active on behalf of Earth, both for its well-being and ours as human beings. Rather, Earth would have us look elsewhere in our efforts to encourage others to participate in ecologically responsible activities and lifestyles. To be sure, there needs to be theological justification for such efforts—this commentary itself seeks to do just this. However, preoccupation with the details of theological argument may itself discourage some from being sympathetic to our goals if they feel that their cherished theologies are being attacked. Theological justification is important; it often needs a more patient approach.

A second way Earth might speak to us is on the matter of suffering on behalf of Earth in our advocacy for Earth. As we noted, there is a growing resistance to environmental advocacy throughout the world, often fueled by corporate and/or governmental interests. Those who work for Earth are often viewed as blocking the economic and social advancement of the people and are seen as enemies of corporations and governmental officials. In places where the rule of law is not so stringently enforced, environmental activists are often in danger, often to the point of bodily harm or even death. Yet even in Western democracies, the economic pressures exerted make it dangerous to defend Earth. Sometimes the threats are physical, but often they are professional and economic. With so much money and political power involved in the issues, the stakes are high, and success on behalf of Earth may mean financial harm and loss of influence for those whose actions have been thwarted.

---

[26] William B. Badke, *Project Earth: Preserving the World God Created* (Portland: Multnomah, 1991).

At times, environmental activism has a sort of "trendy" air attached to it. Earth would remind us that at times it can be hard, unpopular, and even costly work, so much so that at times the word "martyr" might be attached to someone who has paid the highest price for activism on behalf of Earth. The word "martyr" has a lot of connotations associated with it, especially in the context of Christian history and tradition. Sensational stories of horrific tortures and suffering are appropriated as exemplars worthy of emulation among the faithful. And frankly, that can be a frightening deterrent. Earth might help us recall that our word "martyr" derives from Greek words connected to the notion of witness. Framed this way, we should see a call to stand and bear witness on behalf of Earth when Earth finds itself without voice in the boardrooms, backrooms, and courtrooms, testifying on its behalf when it is threatened without representation. Our testimony may take many forms, and indeed, there may sometimes be costs associated with such witness. But a sense of perspective is necessary here. Even being a Christian has historically cost believers their lives, but it is a relatively small percentage who have paid this ultimate price. There may be other costs as well, but the cost of indifference will prove even higher, surely in the long run, perhaps in the short term as well. And we must always remember, as human beings, we, too, have a vested interest in Earth's well-being.

The Goal of Creation: The Dwelling Place of God (Heb. 12:1–13:25)

Contents

The final exhortation of the author focuses on the negative charge not to follow the model of Esau as well as the positive charge to come to the presence of God now available to those in the new covenant. In essence, this combines the warning to adhere to the new covenant with the argument that the new covenant far exceeds the old covenant. The symbolism of Mount Sinai and Mount Zion, while effective in achieving the author's rhetorical objectives, creates the impression that the present creation is destined for destruction at the end of the age. Our ecological reading will argue that the comparison of mountains actually suggests not the destruction of the present order but its transformation into the dwelling place of God, a dwelling in which those under the new covenant both presently and proleptically inhabit. This is the climax of the argument, both for the author's purposes and ours.

Structure

12:1-17: One last warning to hold fast to the new covenant
12:18-29: The New Jerusalem, God's dwelling place
13:1-25: Becoming the community of the New Jerusalem

Analysis

One Last Warning to Hold Fast to the New Covenant (12:1-17)

This first section of Hebrews 12 is a mixture of exhortation and dire warning, as we have seen on numerous occasions throughout the letter. The chapter begins with a call to follow Jesus with perseverance—it asks readers not only to note how he endured affliction in pursuit of his divine calling, but also to recall the "cloud of witnesses" (v. 1) sketched in ch. 11. In light of the perseverance of those old covenant exemplars of faith and Jesus the "pioneer and perfecter" of our faith (12:2), the readers are to run the race set before them with perseverance.

With v. 3, the author brings to consideration another perspective in the call to perseverance. Acknowledging that the audience has experienced some form of persecution that has, to this point, not resulted in bloodshed for any within the community (v. 4),[1] they are to adopt a perspective that these hardships function as discipline (vv. 5-11). Citing Prov. 3:11-12, the author encourages the audience that the presence of opposition is actual proof positive that God considers them as God's own children. Therefore, they are to consider their hardships as evidence of God's concern and to approach them as helpful on their path to glory.

This encouragement allows for an important consideration in our ecological reading of the letter. In our discussion of Hebrews 11, we saw the mention of Moses, with particular reference to the Exodus. The Exodus narrative presents great difficulties for ecologically oriented readers. The account of the plagues visited upon Pharaoh and the Egyptians depict God as one who shows little regard for the other-than-human creation. Horrific destruction against the animals and the land is perpetrated by God in order to secure the release of the covenant people Israel. To call this account "grey," in Norman Habel's terminology, is an example of gross understatement.

---

[1] Acknowledging that actual episodes of persecution may be in view here, especially in light of the context that compares the lot of the audience with Jesus's own sufferings, William Lane argues that the phrase "to the point of shedding your blood" is perhaps figurative, suggesting that the audience has not done everything they can to resist "sin," which he sees as a periphrasis for the term "sinners" in Heb. 12:3. Thus actual persecution is in view, and the function of this verse is shaming their lack of perseverance in light of their relatively light persecution. See Lane, *Hebrews 9–13*, 417–19.

Elsewhere I have argued that a crucial aspect of God's reclamation project for creation involves the collaboration of God with the other-than-human order of creation to judge human beings for their violations of the God/human/other-than-human triad.[2] With respect to the Exodus narrative, Pharaoh's treatment of the Israelites represented, as we noted in the previous chapter, an obstruction of God's purposes for Israel at a time when Israel had begun to actualize their covenantal responsibility to bring restoration to creation. What appears to be God's capricious violation of Earth's integrity as a subject in its own right is actually the cooperation of creation to bring about the restoration of humanity's rightful role in creation. Rather than viewing the judgments of God against Pharaoh as involving collateral damage inflicted upon the other-than-human components of creation in favor of God's chosen human subjects, Israel, they are again the involvement of these parts of creation in the project to restore human beings to their rightful place as God's priestly co-regents in creation.[3] This understanding is confirmed in the interpretation of the Exodus event in the Wisdom of Solomon.[4] In chs 11–19, the story of the Exodus is reinterpreted to show the work of the personified figure of Lady Wisdom in the history of Israel. Prefacing this reinterpretation is the long depiction of Lady Wisdom as integrally connected with creation as well as the one who enables human beings to relate properly to God. This background enables the author of Wisdom of Solomon to connect the work of Wisdom with the judgment against the Egyptians, articulated in Wis. 11:15-16 in the principle that Egypt was judged for their inappropriate responses to God's creation: "In return for their foolish and wicked thoughts, which led them astray to worship irrational serpents and worthless animals, you sent upon them a multitude of irrational creatures to punish them, so that they might learn that one is punished by the very things by which one sins." The Egyptians, seeing God's works of creation, did not give proper honor to God, but rather made gods out of them. Paul indicts idolatrous humanity of this very sin in Rom. 1:22-23. In Wisdom of Solomon, Egypt represents this inclination of human beings to fail to discern their proper place in the God/human/other-than-human triad.

---

[2] Lamp, *Reading Green*, ch. 5.
[3] For details of this reading, including the opposing ecological readings of Norman Habel and Terence Fretheim, see Lamp, *Reading Green*, 95–101.
[4] For discussion, see Lamp, *Reading Green*, 101–6.

This inclination of the Egyptians did not end with the judgment against Egypt in the Exodus. Human beings continue to neglect their proper role in creation by looking at the other-than-human creation in idolatrous ways. While human beings today may not create idols out of animals or other aspects of the other-than-human order, we do still look to creation in ways that betray our true allegiances. Colossians 3:5 identifies one key human sin with idolatry: greed. In a recent interview, conservationist Jane Goodall identified human greed, whether the individual quest for a more affluent lifestyle or the corporate desire for greater profits, as the greatest impediment to environmental protection.[5] It is abundantly clear that human degradation of the environment in pursuit of unabated human thriving has led to numerous ecological calamities. The convenience of single-use plastic containers has filled the oceans with waste that harms ecosystems and their aquatic inhabitants; burning of fossil fuels has affected global climate patterns so that migration patterns of birds, fish, mammals, and insects have been altered, affecting local ecosystems adversely, not to mention their impacts on sea levels and associated effects on coastal communities; clear cutting of forests has contributed to extinctions and desertification. Viewed from one angle, these are the consequences of human actions against Earth. Viewed from another, the ramifications of these actions on human flourishing may represent Earth's judgment, and by implication God's judgment, against human beings, aimed at driving human beings to change their courses of action with respect to Earth. The hardships that these environmental changes inflict upon human beings, the direct result of human failure to recognize their place in the order of creation, may be the kind of thing that they should regard as discipline from God in order to help them persevere as they strive to fulfill their Adamic vocation in the world on their way to glory. We noted earlier that human beings frequently suffer in conjunction with Earth's suffering of environmental degradation. Perhaps here we should change our perspective, as the author of Hebrews encouraged the readers of this letter to do, such that we view our suffering from environmental degradation as Earth's judgment against us for our idolatry to the end that

---

[5] Sam Baker, "'The Greatest Problem Is Greed,' Says Conservationist Jane Goodall," *Deutsche Welle*, July 12, 2019. Available online: https://www.dw.com/en/environmental-protection-the-biggest-problem-is-greed-says-conservationist-jane-goodall/a-49556942?maca=en-EMail-sharing (accessed August 10, 2019).

we come to repent of our ecological sins and fulfill our Adamic vocation in creation.

Here we may also derive some instruction from the author's caution to the readers of Hebrews not to follow in Esau's path (Heb. 12:16-17). Esau is chided for selling his very birthright to his brother Jacob in exchange for a mere meal. Here the author gives a stark warning that Esau was not able to find repentance even though he sought it bitterly with tears. Read ecologically, this warning is equally stark. There may come a point that human beings may forfeit their birthright to be the priestly co-regents of God in creation such that the birthright is no longer recoverable. Various estimates place the "tipping point" of such issues as global climate change within the lifetimes of those currently living on the planet.[6] At present, there is ample evidence that human degradation of Earth is adversely affecting human life. The only real issue is whether human beings will learn the lessons of these adversities and embrace their vocation to work for the well-being of Earth in time to avert ecological disaster.

The New Jerusalem, God's Dwelling Place (12:18-29)

Though an entire chapter remains in the letter following this passage, I argue that Heb. 12:18-29 functions rhetorically as the climax of the author's argument. The passage is a masterful convergence of emphases that have occurred frequently throughout the letter. It involves comparison between the covenants, warning against apostasy, and encouragement to hold fast in faithfulness. The organizing principle of the section is a comparison between two mountains that not only represent the old and new covenants but also speak to the eschatological reality toward which God is moving both creation and the covenant community. Following the warning not to follow in the path of Esau (vv. 16-17), the author continues the sense of warning with the imagery associated with Moses's ascent of Mount Sinai in Exodus 19. The author's quotation of Exod. 19:12 in Heb. 12:20 summarizes the scene of fear and dread that accompanied Moses's visitation with God on the mountain: should even

---

[6] For example, The National Academies of Science, Engineering, and Medicine, *Understanding Earth's Deep Past: Lessons for Our Climate Future* (Washington DC: The National Academies Press, 2011), 63–80.

an animal touch the mountain, it must be killed. Thunder, lightning, fire, gloom, a trumpet sound, and a tempest create a scene of absolute horror that forbade all but Moses to approach, causing the people to beg that the scene come to an end.

The strong adversative conjunction in v. 22 (ἀλλά) signals the beginning of a comparison. If Mount Sinai represents the previous covenantal administration, then Mount Zion would represent the new covenant. The descriptors of Mount Zion contrast with the scene of fear associated with Mount Sinai. This mountain is identified as the "city of the living God," indeed, the "heavenly Jerusalem" itself, populated with "innumerable angels in festal gathering," the "firstborn who are enrolled in heaven," "God the judge of all," the "spirits of the righteous made perfect," and even "Jesus, the mediator of a new covenant" whose blood establishes the new covenant (vv. 22-24).

In the previous chapter, we discussed this passage as belonging to a trajectory within Hebrews that we characterized as one of the "transience of the created order." By this we meant that Earth is depicted as temporary in nature, to be discarded when the eschatological promises of God are culminated. Two features within 12:18-29 are noted in this identification. First, the reality of Mount Sinai was one that could be "touched" (v. 18). The verb ψηλαφάω denotes the act of touching by "feeling and handling," a tactile experience.[7] The implied contrast with Mount Zion is that the heavenly city is one that cannot be touched, for it is a heavenly reality. Indeed, this detail is often cited as evidence that the fate of the present order of creation, particularly in its physicality, is destined for destruction, for in vv. 25-29, the author of Hebrews cites and adapts Hag. 2:6 to indicate that there will come a judgment that will shake both earth and heaven, in which the "removal of what is shaken" (v. 27), namely, "created things," will take place, leaving only those things that cannot be shaken. Some scholars have seen this formulation as deriving from a dualism that, in its present context, indicates the destruction of that part of creation that is physical, leaving behind only that which is heavenly in nature.[8] Such an understanding would represent a bias against Earth that favors a heavenly destiny for those who persevere faithfully to the end.

---

[7] BDAG, 1097.
[8] Attridge, *The Epistle to the Hebrews*, 381; Ellingworth, *The Epistle to the Hebrews*, 685–89; Schenck, *Cosmology and Eschatology in Hebrews*, 109, 139–42.

I offered a brief response to this bias in my earlier volume, one that I will recall here and expand upon.[9] First, we must entertain George Wesley Buchanan's charge that such a view reflects more upon the interpreter's theological convictions than upon what the writer of Hebrews is actually arguing.[10] William Lane argues that the nature of the judgment characterized here by "shaking" is one that in Jewish thought occurs temporally within history, with the net effect of its mention in Hebrews being that in view is the fate of the community of believers and not the physical creation itself.[11] The question here is whether the community will be able to endure the final judgment, not whether the physical cosmos will face annihilation in the eschaton. Jon Laansma goes further, arguing that the reference to shaking has in view the removal of those things that characterize the present order of reality such that what remains will be the creation that was originally destined to become the dwelling place of God.[12] Only that in creation that opposes God faces destruction, not the entirety of creation itself.

Another detail frequently overlooked is that the word that is translated in the NRSV as "removal" (v. 27), μετάθεσις, is capable of another translation, one that is preferred in BDAG.[13] Rather than seeing the shaking of heaven and earth resulting in the removal from existence of the earth, what is in view is the "change" or "transformation" of the present order of the created things. What is in store for creation is a "sorting out and putting right" of those things in creation that are contrary to God, and it is this order within creation that will be changed from the temporariness of corruption into the permanence of the kingdom of God.[14]

The foregoing has been a rehearsal of what I covered in the earlier volume. To this I would now like to add two observations. The first is the mention in v. 27 of the "created things" (πεποιημένων). In the structure of the author's argument, the created things seem to refer to the things that are shaken. Such a conclusion is not unfounded, for in the LXX of Gen. 1:1, the same verb, ποιέω, is used to describe God's creative activity with respect to the heavens and the

---

[9] Lamp, *The Greening of Hebrews?*, 79–80.
[10] George W. Buchanan, *To the Hebrews*, AB (Garden City: Doubleday, 1972), 72–74.
[11] Lane, *Hebrews 9–13*, 480–83. Lane cites in support Isa. 13:1-22; Hag. 2:6; *Jub.* 1:29; *1 En.* 45:1; *Sib. Or.* 3:675-80; and *2 Bar.* 32:1; 59:3.
[12] Laansma, "Hidden Stories in Hebrews: Cosmology and Theology," 14.
[13] BDAG, 639.
[14] Witherington, III, *Letters and Homilies for Jewish Christians*, 346.

earth. But as we have argued, what the author of Hebrews may be describing here is the transformation of these things, not their obliteration. Perhaps what we may be encountering here is not a reference to the things that God has created, but rather to the things that have been created by human beings. In other words, the created things in Heb. 12:27 function in the same semantic space as those things "made by human hands" (χειροποίητος). The author has used this designation in 9:11 and 24 to refer to the sanctuary made by human hands. This sanctuary has been surpassed by the Son in the new covenant. The things made by human hands, in this case the sanctuary, had to be replaced with their heavenly realities. In 12:27, a similar idea may be present.

Perhaps an event in Daniel 2 sheds light here. The Babylonian king, Nebuchadnezzar, has a terrifying dream that only Daniel is able to interpret. In vv. 31-35, Daniel recounts for the king the content of the king's dream:

> You were looking, O king, and lo! there was a great statue. This statue was huge, its brilliance extraordinary; it was standing before you, and its appearance was frightening. The head of that statue was of fine gold, its chest and arms of silver, its middle and thighs of bronze, its legs of iron, its feet partly of iron and partly of clay.
>
> As you looked on, a stone was cut out, not by human hands, and it struck the statue on its feet of iron and clay and broke them in pieces. Then the iron, the clay, the bronze, the silver, and the gold, were all broken in pieces and became like the chaff of the summer threshing floors; and the wind carried them away, so that not a trace of them could be found. But the stone that struck the statue became a great mountain and filled the whole earth.

Verses 36-45 consist of Daniel's interpretation of the dream in which Daniel explains that the statue represents Nebuchadnezzar, who will be destroyed with a stone that was cut "without hands," and that Nebuchadnezzar's kingdom will be replaced by a succession of inferior kingdoms until such time that God will establish a kingdom that will never be destroyed, having destroyed all its predecessors (v. 44). This kingdom is the mountain that results when the statue and all its parts have been rendered to dust (v. 45).

The similarities to Heb. 12:25-29 are striking. The mention of an enduring mountain representing the kingdom of God that results when the contents of the earth have been shaken is reflected in the Daniel account. In the account in Daniel, there is every indication that what was obliterated was not the earth, but those things in the present, corrupted order of creation that stand as obstacles for the full realization of God's kingdom on earth. What remains is what God (re)creates: the earth rid of human corruption. Indeed, if one looks quite literally at what has not been created by human hands, the whole of creation would certainly qualify. The end result of God's shaking of the earth, as the writer of Hebrews calls it, would be an earth that finally realizes its destiny to be the place where God rules in God's kingdom.

As for the second observation I would like to add, what is overlooked in surveys of this passage is the significance of the statement in v. 22 where the author asserts by way of encouragement that the readers "have come to" Mount Zion. Here the verb translated "have come" is the perfect tense of the verb προσέρχομαι. Bypassing the technicalities of the linguistic discussions of the perfect tense in the Greek verb system,[15] the author's choice of the perfect tense at least seems to suggest that the readers currently, in some proleptic sense, already experience the realities of the heavenly Jerusalem. While the fullness of this experience awaits the climax of the ages with the return of Jesus, it is already the present experience of the community. They already come to Mount Zion and its concomitant realities when they worship the risen Jesus. The crucial point here from an ecological perspective is that this present reality is taking place within the physical creation. It does not need to await the culmination at the end of the age for it to be the real experience of the people. This shows, in an anticipatory way, that the realities of the eschatological kingdom of God, a kingdom that cannot be shaken, are already present in the world. If these rather liturgical realities are already present, indicating that God is already present in the world as a foretaste of the final destiny of the created order, then the people of God need to live in anticipation of their final realization. From an ecological perspective, this entails the people of God fulfilling their Adamic vocation to serve and protect Earth in anticipation of its final destiny of becoming the dwelling place of God. Yes, God will one day finish the reclamation project and remove those things that obstruct the realization of this destiny, but in the meantime, the people of God must, in every way, embody this reality in the world and fulfill the purpose for which they were created and recreated. This is, in N. T. Wright's idiom, "eschatology in the process of realization."

---

[15] For a brief discussion, see Constantine R. Campbell, *Advances in the Study of Greek: New Insights for Reading the Greek New Testament* (Grand Rapids: Zondervan, 2015), 117–19.

Becoming the Community of the New Jerusalem (13:1-25)

With Hebrews 13, we come to a point in the epistle where the content of the letter shows some semblance to the specific construction of an epistle. To this point, what we have seen may be characterized as a homily or a piece of Greco-Roman rhetoric, and indeed scholars have noted this. But in ch. 13, there is the stream of various specific exhortations and what appears to be an epistolary conclusion (vv. 18-25). The exhortations that begin the chapter are not what one would necessarily expect to encounter given what has gone before. They are rather mundane and do not appear to derive from the specific preceding argumentation. However, there is a section, vv. 9-16, in which themes explicated throughout the letter emerge and occupy the author's attention. Verse 9 includes mention of food laws that are ineffectual for those who observe them; v. 10 asserts there is an altar in the new covenant before which it is inappropriate for those who officiate in the old covenant to serve; vv. 11-13 connect the sacrifice of Jesus outside the city with those animals who were sacrificed on the Day of Atonement and whose bodies were burned outside the city, showing Jesus to be the intention of the former observance;[16] v. 14 makes mention of the city for which we look that is not to be found in this order of creation. Verses 15 and 16 draw attention to ways in which worship now appropriate in the new covenant does not involve the ritual performance of the old covenant. As we have had opportunity to examine these themes earlier, a rehearsal of their ecological significance need not occupy us here. Rather, we will take another approach at this point.

In my earlier volume, I took v. 10 and the mention of the altar as a point of departure to introduce the Eucharist into the discussion.[17] Here, I wish to look at the assertions made in vv. 9-14 as merely reminders of what we have already discussed as explanatory of what is probably the most well-known verse in the entire letter, v. 8, which reads, "Jesus Christ is the same yesterday and today

---

[16] There is an irony here. The author notes that those who ministered under the old covenant were not permitted to eat at the altar constituted by the sacrifice of Jesus. But those who performed the sacrifices on the Day of Atonement, which is in view here, were also not permitted to eat of the sacrifices they performed. Thus the animals' bodies must be burned outside the city. I discuss this in Lamp, *The Letter to the Hebrews*, 129–30.

[17] Lamp, *The Greening of Hebrews?*, 85–100.

and forever," and as anticipatory of vv. 15-16, which speak of the sacrifices appropriate in light of Jesus's sacrifice.

Verse 8 functions for the author to affirm the value in observing the life of those who went before the first audience as a model for their own perseverance in faith. Ecologically the verse functions to argue that looking to what we have observed of Jesus as a rubric for our own ecological pilgrimage stands for all time. The Son is God's agent of creation, the one who was not only involved in its origin but who also sustains the creation by his powerful word and who is the heir of all that he was involved in creating (1:2-3). In his incarnation, Jesus recapitulated in himself as the Second Adam all that was mandated of the first human beings in their priestly co-regency with God to extend God's benevolent reign into all that God has created through the Son and bequeathed to him as an inheritance in anticipation of the time God would inhabit creation. In his recapitulation of the Adamic vocation Jesus has entrusted this vocation back to those who participate in the new covenant. Hebrews 13:9-14, then, remind us of the specific aspects of the totality of the Son's work to bring about the renewal of this vocation for us.

Verses 15 and 16 list the types of sacrifices that are appropriate for the adherents of the new covenant. A sacrifice of praise that confesses God's name (v. 15) and good works and sharing what we have (v. 16) describe this sacrifice. First Peter 2:5 describes the people of God as living stones crafted into a house of worship, a priesthood that offers spiritual sacrifices to God through Christ. If we were to inquire what kinds of sacrifices of praise would be worthy of God, ecologically speaking, what would they look like? First Peter 2:9 speaks of the proclamation of this priesthood, consisting of the proclamation of God's mighty acts. From an ecological point of view, this might entail the proclamation of what God has accomplished in Christ, namely, the reestablishment of the Adamic priestly vocation to the people of God. Whatever else the authors of Hebrews and 1 Peter may have had in mind, if the priesthood established among the people of God involves the reclamation of the Adamic priesthood in creation, then the praises and proclamations would be for the renewal of this priesthood. We praise God that God has set things right and given Earth the priesthood that was to have been from the beginning.

But praises and spiritual sacrifices are not in themselves enough. They must be substantiated by good works and sharing (Heb. 13:16). Human beings can speak of what God has done; true faithful perseverance toward glory must consist of action commensurate with confession. This is the real meaning of the word "faith" in Hebrews. The two specific items mentioned, good works and sharing, are especially appropriate in an ecological context. What kinds of good works would be appropriate to express the Adamic vocation toward creation? Any number come to mind: recycling, reducing consumptive practices, minimizing carbon footprints, adopting a Sabbath lifestyle.[18] Several popular level studies extol the benefits of caring for creation for the spiritual life of followers of Jesus.[19] Many concrete expressions of the Adamic vocation to care for creation are well within reach of the average person.

Sharing with each other is an expression that involves a reorientation for many of the faithful. Especially in the Western world, individual well-being is often advanced as a virtue, with emphasis on communal well-being viewed with suspicion. As we noted earlier, Col. 3:5 equates greed with idolatry. Understanding that the world's resources are gifts of God to all human beings, to be used with proper respect both for other persons and for Earth, is the critical factor for achieving what the author of Hebrews exhorts here. Human beings were originally commissioned to see to it that Earth is able to attain to its destiny to become the dwelling place of God. Unbridled exploitation of Earth's resources for the satiation of the few at the expense of the many denies both Earth and human beings full opportunity to thrive. Human beings must come to a place where they realize that by their actions they may affect, for good or ill, the well-being of all things. Realizing that we are called to seek the benefits of others and of all of creation, we must come to a place where we are not seeking to satisfy every personal desire, but know our place within creation and seek its well-being.

Hearing Earth

In this final opportunity to hear Earth's voice, we hear a simple call. Earth calls out for us to accept the mantle of the Adamic vocation for which we were created and to perform sacrifices of good works on behalf of Earth as we anticipate our

---

[18] For discussions of how reclaiming the Sabbath as a pattern for living in the world affects ecological matters, see A. J. Swoboda, *Subversive Sabbath: The Surprising Power of Rest in a Nonstop World* (Grand Rapids: Brazos, 2018). Chapters 7–9 deal with matters of Sabbath with respect to the order of creation, the land, and the animals, respectively.

[19] For example, J. Matthew Sleeth, *Serving God, Saving the Planet: A Call to Care for Creation and Your Soul* (Grand Rapids: Zondervan, 2012).

joint liberation from corruption and death. If indeed the heavenly Jerusalem is already in our midst in some way, it behooves us to conduct ourselves in a way that accords with this reality. If we already experience the divine presence that will characterize creation once God comes to inhabit it along with God's image-bearing human priestly co-regents, then we must perform that priestly ministry on behalf of Earth so that Earth may proceed toward the hope that awaits it and us. In a series of dialogues with N. T. Wright, John Dominic Crossan coined the phrase "collaborative eschatology" to refer to the mission of God's people to establish in an anticipatory way the new creation toward which God is moving history.[20] The ecological expression of this future would be those good works performed as priestly sacrifice on behalf of Earth that lets it, too, experience a foretaste of the glory awaiting it, just as Earth's human constituents await this glory with eagerness. Those renewed in the Adamic vocation must take upon themselves the work of Earth's healing, seeing in Earth a co-recipient of God's reclamation project. It is for its own freedom to pursue its destiny to be God's dwelling place that Earth cries out in desperation and hope for human beings to cease being the cause of Earth's ecological suffering and to be the agents of its healing. The call is prophetic, and the substance of this call is really that simple.

---

[20] Robert B. Stewart, John Dominic Crossan, and N. T. Wright, *The Resurrection of Jesus: John Dominic Crossan and N.T. Wright in Dialogue* (Minneapolis: Fortress Press, 2006).

12

Conclusion

Now at the conclusion of our reflections, some summary comments are necessary. By this time it should not need reiteration that what has preceded does not aspire to the contemporary understanding of what constitutes a biblical commentary. I have made only the most cursory comments on what the author of Hebrews was doing at each stage of the argument, and then only to establish some contours for how I would then engage this material from an ecological perspective.

Also, at this point, it bears summarizing what is meant by the phrase "ecological perspective." My approach here was not to address a whole catalog of contemporary issues, though some general examples are cited. I chose this approach for a couple of reasons. First, the news cycle on environmental issues changes rapidly. By the time the volume came to print, in all likelihood specifics of any issues I chose to examine may have changed, and indeed, some issues may have become either more or less crucial in the public estimation than they would have appeared in the commentary. Such matters of detail would be a potential distraction. Second, issue-specific discussions of any depth would have seemed, in my opinion, to be out of character for the perspective of the author and the shape of the letter. Hebrews seems more interested in "big picture" theological matters than specific exigences. Indeed, the specific reasons the author chose to write are frustratingly missing. Rather, what we find in Hebrews is a sustained exhortation for readers to remain faithful to the covenant instituted by Jesus—an exhortation that is heavily grounded in substantial theological reflection on just how this covenant differs from the previous one and why it is the one to which readers must adhere. So my decision was to adopt an ecological perspective that itself was more theologically framed than issue driven. As stated earlier, I am highly sympathetic with Ernst Conradie's contention that a true ecological reading of the Bible must proceed from a rigorous attempt to conceive of the cardinal doctrines of Christianity in ecological terms. My desire was to contribute, within the parameters of a commentary in this series, to such a rearticulation.

The tack taken here was to follow the lead of the author, to allow the actual letter to develop its argument, and then to use aspects of the argument as points of departure for ecological considerations. Kenneth Schenck has argued that the argument of Hebrews proceeds from a series of narratives that underlie the rhetoric of the letter.[1] The story of human redemption is reflected in the letter, an obvious anthropocentric bias. From this story, then, my ecological reading proceeded through either subversion of the author's own language or extension of the scope of the author's argument beyond simply human subjects to Earth. The result was its own story of redemption, beginning with the creation of the cosmos through the Son and his incarnation. As the story continued, we saw the precise mechanism for this redemption was the recapitulation of the Adamic vocation as presented in Gen. 1:26-28 and 2:15. This vocation, succinctly stated, was to be the priestly co-regents with God to extend God's benevolent rule throughout all creation with the goal being the establishment of creation as the dwelling place of God with all that God had created. The presentation of Jesus as the true high priest who reigns over creation from the right hand of God becomes the rubric for those who adhere to the new covenant to resume the Adamic vocation now reclaimed through the Second Adam. The result is the healing of the ruptures introduced by human sin into the God/human/other-than-human triad.

Ethical programs and plans of action, of course, have their place in the healing of Earth. It has been my experience, however, that over time, such engagements originating from a Christian orientation require a biblical-theological justification, especially when attention is directed toward enlisting those not previously committed to ecological expressions of their faith. Given the richness of christological reflection in Hebrews, the letter serves as a significant test case for whether an extended ecological engagement with it may produce insights that inform an ecological rearticulation of christology that, in turn, would spur the faithful on to action on behalf of Earth.

---

[1] Kenneth L. Schenck, *Understanding the Book of Hebrews: The Story behind the Sermon* (Louisville: Westminster John Knox, 2005). A similar approach is employed by Matthew C. Easter, *Faith and the Faithfulness of Jesus in Hebrews*, Society for New Testament Monograph Series 160 (New York: Cambridge University Press, 2014).

I have no delusions that much of what was presented in this study would have occurred to the author of Hebrews when penning this letter to its original recipients. As with many approaches to reading the Bible that may be termed "ideological," the question is whether there is a way to begin with issues pertinent in a modern context as we approach an ancient text not directly concerned with such issues. Does the Bible have anything to contribute to current discussions when present conditions were simply not evident in the ancient context? One could certainly take the approach of *The Green Bible* and simply highlight in a green font texts that directly speak of the natural world.[2] Norman Habel critiques this rather naive approach as inadequate.[3] What is required is an approach that acknowledges that ecological concerns were not prioritized in the Bible but still seeks to engage the Bible critically for its potential contributions to the current ecological crises facing the world. The Earth Bible Project and the Exeter Group each attempt in their own ways to embark upon this task. In my own way, I have attempted to contribute to the methodological conversation in the volume *Reading Green*, concluding that study with some guidelines for approaching the Bible from a position of ecological concern.[4] It is my hope that this commentary on the letter to the Hebrews has contributed in some small way to interrogating the Bible for its own contribution to this effort.

---

[2] *The Green Bible* (New York: HarperOne, 2008).
[3] Norman Habel, "When Earth Reads *The Green Bible*," *Nature and Culture* 3 (2009): 421–22.
[4] Lamp, *Reading Green*, 133–34.

Bibliography

"164 Activists Were Killed Defending Land and Water Last Year." *YaleEnvironment360*. July 30, 2019. Available online: https://e360.yale.edu/digest/164-activists-were-killed-defending-land-and-water-last-year. Accessed August 3, 2019.

Andreopoulos, Andreas. "'All in All' in the Byzantine Anaphora and the Eschatological Mystagogy of Maximos the Confessor." In *Studia Patristica*, vol. 68, edited by Markus Vinzent, 303–12. Leuven: Peeters, 2013.

Attridge, Harold W. *The Epistle to the Hebrews*. Hermeneia. Philadelphia: Fortress, 1989.

Badke, William B. *Project Earth: Preserving the World God Created*. Portland: Multnomah, 1991.

Baker, Sam. "'The Greatest Problem Is Greed,' Says Conservationist Jane Goodall." *Deutsche Welle*. July 12, 2019. Available online: https://www.dw.com/en/environmental-protection-the-biggest-problem-is-greed-says-conservationist-jane-goodall/a-49556942. Accessed August 10, 2019.

Barker, Margaret. *Creation: A Biblical Vision for the Environment*. London and New York: T&T Clark, 2010.

Bartholomew I. *On Earth as in Heaven: Ecological Vision and Initiatives of Ecumenical Patriarch Bartholomew*, edited by John Chryssavgis. New York: Fordham University Press, 2012.

Bauckham, Richard. *The Bible and Ecology: Rediscovering the Community of Creation*. Waco: Baylor University Press, 2010.

Bauckham, Richard. *Living with Other Creatures: Green Exegesis and Theology*. Waco: Baylor University Press, 2011.

Bauer, Walter, Frederick W. Danker, W. F. Arndt, and F. W. Gingrich. *Greek-English Lexicon of the New Testament and Other Early Christian Literature*, 3rd ed. Chicago: University of Chicago Press, 2000.

Beale, Gregory K. *The Temple and the Church's Mission: A Biblical Theology of the Dwelling Place of God*. Downers Grove: IVP, 2004.

Bordeianu, Radu. "Maximus and Ecology: The Relevance of Maximus the Confessor's Theology of Creation for the Present Ecological Crisis." *The Downside Review* 127 (2009): 103–26.

Brown, William P. *The Seven Pillars of Creation: The Bible, Science, and the Ecology of Wisdom*. Oxford: Oxford University Press, 2010.

Buchanan, George W. *To the Hebrews*. Anchor Bible. Garden City: Doubleday, 1972.

Butt, Nathalie, Frances Lambrick, Mary Menton, and Anna Renwick. "The Supply Chain of Violence." *Nature Sustainability* 2 (2019): 742–47.

Byrne, Brendan. "An Ecological Reading of Rom. 8.19–22: Possibilities and Hesitations." In *Ecological Hermeneutics: Biblical, Historical, and Theological Perspectives*, edited by David G. Horrell, Cherryl Hunt, Christopher Southgate, and Francesca Stavrakopoulou, 83–93. London and New York: T&T Clark, 2010.

Cadwallader, Alan H. "Earth as Host or Stranger? Reading Hebrews 11 from Diasporan Experience." In *The Earth Story in the New Testament*, edited by Norman C. Habel and Vicky Balabanski, 148–65. London: Sheffield Academic, 2002.

Campbell, Constantine R. *Advances in the Study of Greek: New Insights for Reading the Greek New Testament*. Grand Rapids: Zondervan, 2015.

Carson, Rachel. *Silent Spring*. Boston: Houghton Mifflin, 1962.

Chrysostom, John. *The Divine Liturgy of Our Father among the Saints, John Chrysostom*, translated by Seraphim Dedes. Columbia: Newrome Press, 1995.

Chryssavgis, John. "Icons, Liturgy, Saints: Ecological Insights from Orthodox Spirituality." *International Review of Mission* 99 (2010): 181–89.

Chryssavgis, John, and Bruce V. Foltz, eds. *Toward an Ecology of Transfiguration: Orthodox Christian Perspectives on Environment, Nature, and Creation*. New York: Fordham University Press, 2013.

Conradie, Ernst M. "What on Earth Is an Ecological Hermeneutics? Some Broad Parameters." In *Ecological Hermeneutics: Biblical, Historical, and Theological Perspectives*, edited by David G. Horrell, Cherryl Hunt, Christopher Southgate, and Francesca Stavrakopoulou, 295–313. London and New York: T&T Clark, 2010.

Craddock, Fred B. *Hebrews*. New Interpreter's Bible. Nashville: Abingdon, 1998.

Easter, Matthew C. *Faith and the Faithfulness of Jesus in Hebrews*. Society for New Testament Monograph Series 160. New York: Cambridge University Press, 2014.

Edwards, Denis. *Ecology at the Heart of Faith*. Maryknoll: Orbis, 2006.

Edwards, Lucy E. "What Is the Anthropocene?" *EOS: Earth and Space Science News*. November 30, 2015. Available online: https://eos.org/opinions/what-is-the-anthropocene. Accessed July 18, 2019.

Ellingworth, Paul. *The Epistle to the Hebrews*. New International Greek Testament Commentary. Grand Rapids: Eerdmans, 1991.

"Environmentalists at Risk: An E360 Series." *YaleEnvironment360*. February 27, 2017, March 7, 2017, March 13, 2017. Available online: https://e360.yale.edu/series/environmentalists-at-risk. Accessed August 3, 2019.

Fewster, Gregory P. *Creation Language in Romans 8: A Study in Monosemy*. Linguistic Biblical Studies 8. Leiden: Brill, 2013.

Francis (Pope). *Laudato Si': On Care for Our Common Home*. Vatican City: Libreria Editrice Vaticana, 2015.

Fretheim, Terence. *The Book of Genesis*. New Interpreter's Bible. Nashville: Abingdon, 1994.

Fretheim, Terence. *Exodus*. Interpretation. Louisville: John Knox, 1991.

Green Bible, The. New York: HarperOne, 2008.

Green, Chris E. W. *Toward a Pentecostal Theology of the Lord's Supper: Foretasting the Kingdom*. Cleveland: CPT Press, 2012.

Habel, Norman C. "When Earth Reads *The Green Bible*." *Nature and Culture* 3 (2009): 421–22.

Habel, Norman C. *The Birth, the Curse and the Greening of Earth: An Ecological Reading of Genesis 1–11*. The Earth Bible Commentary 1. Sheffield: Sheffield Phoenix, 2011.

Habel, Norman C. *An Inconvenient Text: Is a Green Reading of the Bible Possible?* Hindmarsh: ATF, 2009.

Habel, Norman C. *The Land Is Mine: Six Biblical Images*. Minneapolis: Fortress, 1995.

Habel, Norman C., David Rhoads, and H. Paul Santmire. *The Season of Creation: A Preaching Commentary*. Minneapolis: Fortress, 2011.

Hatzidakis, Emmanuel. *The Heavenly Banquet: Understanding the Divine Liturgy*. Chicago: Orthodox Witness, 2010.

Heen, Erik M., and Philip D. W. Krey, eds. *Hebrews*. Ancient Christian Commentary on Scripture. Downers Grove: IVP Academic, 2005.

Hobgood-Oster, Laura. *The Friends We Keep: Unleashing Christianity's Compassion for Animals*. Waco: Baylor University Press, 2010.

Hogue, Michael S. *The Tangled Bank: Toward an Ecotheological Ethics of Responsible Participation*. Eugene: Pickwick, 2008.

Horrell, David G., Cherryl Hunt, and Christopher Southgate. *Greening Paul: Rereading the Apostle in an Age of Ecological Crisis*. Waco: Baylor University Press, 2010.

Huddleston, Jonathan. *Eschatology in Genesis*. Forschungen zum Alten Testament 2. Reihe. Tübingen: Mohr Siebeck, 2012.

Huff, Peter A. "Calvin and the Beasts: Animals in John Calvin's Theological Discourse." *Journal of the Evangelical Theological Society* 42 (March 1999): 67–75.

Imhoff, Daniel, ed. *CAFO: The Tragedy of Industrial Animal Factories*. Los Angeles: Foundation for Deep Ecology, 2010.

Intergovernmental Panel on Climate Change. *Climate Change: Synthesis Report. Contribution of Working Groups I, II and III to the Fifth Assessment Report of the*

Intergovernmental Panel on Climate Change, edited by R. K. Pachauri and L. A. Meyer. Geneva, Switzerland: IPCC, 2014.

Intergovernmental Panel on Climate Change. *Global Warming of 1.5°C. An IPCC Special Report on the Impacts of Global Warming of 1.5°C above Pre-Industrial Levels and Related Global Greenhouse Gas Emission Pathways, in the Context of Strengthening the Global Response to the Threat of Climate Change, Sustainable Development, and Efforts to Eradicate Poverty*, edited by Valérie Masson-Delmotte, Panmao Zhai, Hans-Otto Pörtner, Debra Roberts, Jim Skea, Priyadarshi R. Shukla, Anna Pirani, Wilfran Moufouma-Okia, Clotilde Péan, Roz Pidcock, Sarah Connors, J. B. Robin Matthews, Yang Chen, Xiao Zhou, Melissa I. Gomis, Elisabeth Lonnoy, Tom Maycock, Melinda Tignor, and Tim Waterfield. Geneva, Switzerland: World Meteorological Organization, 2018.

Jewett, Robert. *Letter to Pilgrims: A Commentary on the Epistle to the Hebrews*. New York: Pilgrim Press, 1981.

Johnson, Luke Timothy. *Hebrews*. New Testament Library. Louisville: Westminster John Knox, 2006.

Koester, Craig R. *Hebrews*. Anchor Bible. New York: Doubleday, 2001.

Laansma, Jon. "Hidden Stories in Hebrews: Cosmology and Theology." In *A Cloud of Witnesses: The Theology of Hebrews in Its Ancient Contexts*, edited by Richard Bauckham, Daniel Driver, Trevor Hart, and Nathan MacDonald, 9–18. London: T&T Clark, 2008.

Lamp, Jeffrey S. *First Corinthians 1-4 in Light of Jewish Wisdom Traditions: Christ, Wisdom, and Spirituality*. Lampeter: Edwin Mellen, 2000.

Lamp, Jeffrey S. *The Greening of Hebrews? Ecological Readings in the Letter to the Hebrews*. Eugene: Pickwick, 2012.

Lamp, Jeffrey S. *The Letter to the Hebrews: A Centre for Pentecostal Theology Bible Study*. Cleveland: CPT Press, 2017.

Lamp, Jeffrey S. *Reading Green: Tactical Considerations for Reading the Bible Ecologically*. New York: Peter Lang, 2017.

Lamp, Jeffrey S. "Wisdom Pneumatology and the Creative Spirit: The Book of Wisdom and the Trinitarian Act of Creation." *Spiritus: ORU Journal of Theology* 2 (2017): 39–56.

Lane, William L. *Hebrews 1-8*. Word Biblical Commentary. Dallas: Word, 1991.

Lane, William L. *Hebrews 9-13*. Word Biblical Commentary. Dallas: Word, 1991.

Linzey, Andrew. *Animal Gospel: Christian Faith as if Animals Mattered*. Louisville: Westminster John Knox, 2000.

Linzey, Andrew. "C. S. Lewis' Theology of Animals." *Anglican Theological Review* 80 (Winter 1998): 60–81.

Lodahl, Michael. *God of Nature and of Grace: Reading the World in a Wesleyan Way*. Nashville: Kingswood, 2003.

Louth, Andrew. "Man and Cosmos in St. Maximus the Confessor." In *Toward an Ecology of Transfiguration: Orthodox Christian Perspectives on Environment, Nature, and Creation*, edited by John Chryssavgis and Bruce V. Foltz, 59–71. New York: Fordham University Press, 2013.

Lowe, Ben. *Green Revolution: Coming Together to Care for Creation*. Downers Grove: IVP, 2009.

McFague, Sally. *The Body of God: An Ecological Theology*. Minneapolis: Augsburg Fortress, 1993.

Michaels, J. Ramsey. "The Redemption of Our Body: The Riddle of Romans 8:19–22." In *Romans and the People of God: Essays in Honor of Gordon D. Fee on the Occasion of His 65th Birthday*, edited by Sven K. Soderlund and N. T. Wright, 92–114. Grand Rapids: Eerdmans, 1999.

Middleton, Richard. *The Liberating Image: The* Imago Dei *in Genesis 1*. Grand Rapids: Brazos, 2005.

Moltmann, Jürgen. *God in Creation: A New Theology of Creation and the Spirit of God*, translated by Margaret Kohl. San Francisco: Harper & Row, 1985.

Morgan, Jonathan. "Sacrifice in Leviticus: Eco-Friendly Ritual or Unholy Waste?" In *Ecological Hermeneutics: Biblical, Historical, and Theological Perspectives*, edited by David G. Horrell, Cherryl Hunt, Christopher Southgate, and Francesca Stavrakopoulou, 32–45. London and New York: T&T Clark, 2010.

Munteanu, Daniel. "Cosmic Liturgy: The Theological Dignity of Creation as a Basis of an Orthodox Ecotheology." *International Journal of Public Theology* 4 (2010): 332–44.

National Academies of Science, Engineering, and Medicine, The. *Understanding Earth's Deep Past: Lessons for Our Climate Future*. Washington, DC: The National Academies Press, 2011.

Northcott, Michael S. *The Environment and Christian Ethics*. New Studies in Christian Ethics. Cambridge: Cambridge University Press, 1996.

Painter, John. "Review of Margaret Barker, *Creation: A Biblical Vision for the Environment*." *Review of Biblical Literature* (2011). Available online: http://www.bookreviews.org. Accessed July 18, 2019.

Peeler, Amy L. B. "Review of Jeffrey S. Lamp, *The Greening of Hebrews? Ecological Readings in the Letter to the Hebrews*." *Review of Biblical Literature* (2014). Available online: http://www.bookreviews.org. Accessed July 18, 2019.

Richter, Sandra. "Environmental Law in Deuteronomy: One Lens on a Biblical Theology for Creation Care." *Bulletin for Biblical Research* 20 (2010): 355–76.

Schaeffer, Francis A. *Pollution and the Death of Man: The Christian View of Ecology*. Wheaton: Tyndale House, 1970.

Schenck, Kenneth L. *Cosmology and Eschatology in Hebrews: The Settings of the Sacrifice*. Cambridge: Cambridge University Press, 2007.

Schenck, Kenneth L. *Understanding the Book of Hebrews: The Story behind the Sermon*. Louisville: Westminster John Knox, 2005.

Schmemann, Alexander. *For the Life of the World: Sacraments and Orthodoxy*. Crestwood: St. Vladimir's Seminary Press, 1973.

Silva, Moises, ed. *New International Dictionary of New Testament Theology and Exegesis*. Grand Rapids: Zondervan, 2014.

Sleeth, J. Matthew. *Serving God, Saving the Planet: A Call to Care for Creation and Your Soul*. Grand Rapids: Zondervan, 2012.

Snyder, Howard A., and Joel Scandrett. *Salvation Means Creation Healed: The Ecology of Sin and Grace*. Eugene: Cascade, 2011.

Spencer, Nick, Robert White, and Virginia Vroblesky. *Christianity, Climate Change, and Sustainable Living*. Peabody: Hendrickson, 2009.

Stewart, Robert B., John Dominic Crossan, and N. T. Wright. *The Resurrection of Jesus: John Dominic Crossan and N.T. Wright in Dialogue*. Minneapolis: Fortress Press, 2006.

Stone, Lawson. "Worship as Cherishing Yahweh's World." Paper presented at the annual meeting of the Institute for Biblical Research, New Orleans, LA, November 21, 2009.

Swoboda, Aaron Jason, ed. *Blood Cries Out: Pentecostals, Ecology, and the Groans of Creation*. Eugene: Pickwick, 2014.

Swoboda, Aaron Jason. *Subversive Sabbath: The Surprising Power of Rest in a Nonstop World*. Grand Rapids: Brazos, 2018.

Swoboda, Aaron Jason. *Tongues and Trees: Toward a Pentecostal Ecological Theology*. Journal of Pentecostal Theology Series 40. Blandford Forum: Deo, 2013.

Theokritoff, George. "The Cosmology of the Eucharist." In *Toward an Ecology of Transfiguration: Orthodox Christian Perspectives on Environment, Nature, and Creation*, edited by John Chryssavgis and Bruce V. Foltz, 131–35. New York: Fordham University Press, 2013.

Tonstad, Sigve K. *The Letter to the Romans: Paul among the Ecologists*. Sheffield: Sheffield Phoenix: 2016.

Trainor, Michael. *About Earth's Child: An Ecological Listening to the Gospel of Luke*. Sheffield: Sheffield Phoenix Press, 2012.

Van Dyke, Fred. *Between Heaven and Earth: Christian Perspectives on Environmental Protection*. Santa Barbara: Praeger, 2010.

Wade, Richard. "Towards a Christian Ethics of Animals." *Pacifica* 13 (June 2000): 202–12.

Walton, John H. *Ancient Near Eastern Thought and the Old Testament: Introducing the Conceptual World of the Hebrew Bible*. Grand Rapids: Baker, 2006.

White, Lynn, Jr. "The Historical Roots of Our Ecologic Crisis." *Science* 155 (1967): 1203–7.

Williamson, Ronald. "The Eucharist and the Epistle to the Hebrews." *New Testament Studies* 21 (1975): 300–12.

Wilson, Edward O. *The Creation: An Appeal to Save Life on Earth*. New York: W. W. Norton & Company, 2007.

Witherington, Ben, III. *Letters and Homilies for Jewish Christians*. Downers Grove: IVP Academic, 2007.

Wright, Nicholas Thomas. *Paul and the Faithfulness of God*. Minneapolis: Fortress, 2013.

Wright, Nicholas Thomas. *Simply Christian: Why Christianity Makes Sense*. New York: HarperCollins, 2006.

Wright, Nicholas Thomas. *Surprised by Hope: Rethinking Heaven, the Resurrection, and the Mission of the Church*. New York: HarperOne, 2008.

Zizioulas, John. "Preserving God's Creation: Three Lectures on Ecology and Theology." *King's Theological Review* 12 (1989): 1–5, 41–45; and 13 (1990): 1–5.

Subject Index

Aaron   57–8
Abraham (Abram)   46, 55–6, 62–4, 67, 68–70, 104–8, 109, 110, 111
Adam   12, 19, 34, 35, 36, 42, 50, 57, 60, 62–3, 79, 92, 93, 107, 108, 109, 110
   vocation   12, 27–32, 33–4, 36, 37, 38–9, 45, 52, 55–6, 57, 59, 63–4, 67, 71, 72, 73–5, 76–7, 81, 87, 89, 92–3, 94, 97, 101, 102, 103–4, 107–8, 109, 110, 111, 118–19, 123, 125, 126–7, 130
Adamah   31
agriculture   43, 98–9, 103
angels   15, 20–2, 23, 24, 28, 30–1, 38, 75, 120
animals   31, 46, 87, 88, 89–99, 109–10, 111, 116, 117–18, 124, 126
anthropocentric bias   2–3, 7–11, 16, 24, 31, 33, 41, 46, 52, 63, 64, 130

blood   34, 72–3, 87, 88, 89–97, 102, 109, 111, 116, 120

christology   8, 17, 25, 130
   creational   17, 25–6
climate change   8, 43, 54, 98, 119
concentrated animal feeding operation (CAFO)   98
conscience   36, 74–5, 87–90, 92–4, 111
covenant(al)   5, 11, 12, 22–3, 24, 37, 38, 41, 45–6, 55, 56, 60, 61, 62–4, 76, 80, 82, 87, 96, 99, 101, 102, 104, 107–10, 116, 117, 119, 120, 129
   new   3, 5, 9, 11, 12, 15, 18, 23, 24, 26, 27, 32, 39, 55, 56, 60, 67, 76, 79, 81, 82, 83, 84, 86, 87, 88, 89, 95, 101, 102, 104, 106, 115, 116, 119, 120, 122, 124, 125, 130
   old   5, 9, 10, 12, 15, 16, 20, 35, 37, 38, 39, 55, 57, 58, 67, 76, 79, 81, 82, 83, 85–6, 88, 89–90, 91, 95, 101, 102, 104, 110, 115, 116, 119, 124

Daniel   122
Day of Atonement   10, 89, 95, 124
Divine Liturgy   71–5
dualism   3–6, 16, 80, 120

earth(ly)   4, 5, 12, 13–14, 16, 20, 22, 23, 24, 25, 27, 29, 62, 63, 68, 70, 73, 80, 81, 84–5, 87, 88–9, 90, 92, 93, 96, 106, 107, 108, 109, 120–3
   Earth as subject   2, 3, 4, 7, 8, 9, 10, 12, 13–14, 20–1, 24–6, 29, 31–2, 33, 34, 35–6, 39, 41, 42–3, 45, 50, 52–4, 58, 59–60, 61, 64–5, 70, 71, 72, 73, 75, 76–7, 79, 80, 82, 84–6, 89, 90, 91, 94, 97–9, 103–4, 105, 106, 107, 108, 112–14, 117, 118–19, 120, 123, 125, 126–7, 130
Earth Bible Project   6–8, 14, 131
Eastern Orthodox(y)   30, 64, 71–5, 77
ecological hermeneutics   1–2, 6–9, 58, 64, 70
ecotheology   1, 64–5
environmental justice   8, 98–9
Esau   115, 119
eschatology/eschatological   5, 8, 16, 19, 32, 42, 48–9, 54, 61, 104–7, 112–13, 119–23, 127
Eucharist   70–6, 77, 124
Exeter Group   6, 7–8, 11, 64, 131
Exodus   110–11, 116–18

faith(ful)   3, 5, 6, 11, 12, 18, 22–3, 34, 36, 37, 38–42, 45–8, 51, 52, 55, 56, 59–61, 62, 65, 81, 82, 84, 90, 95, 101, 102–3, 104–5, 110, 114, 116, 119, 120, 125–6, 129, 130
Francis of Assisi   97

Genesis   25, 87, 97
Green Bible, The   6, 131

Holy Spirit   22, 23–4, 39–40, 42–3, 52, 53, 61, 72, 86, 94

Subject Index

incarnation 12, 26, 27, 30, 32, 34, 70, 72, 91–4, 125
Intergovernmental Panel on Climate Change (IPCC) 53–4
Isaac 62–3, 107, 108, 110, 111

Jacob 62–3, 108, 110, 119
Jeremiah 47, 83, 84, 88
Jesus 3, 5, 12, 15, 23, 24, 30–2, 34, 37–8, 45, 49, 50, 53, 55, 56, 57, 58, 60, 62, 67, 73, 74, 75, 80, 82, 85–6, 87, 89, 90, 91, 92–7, 99, 102, 104, 109, 111, 113, 116, 120, 123, 124–5, 126, 129
  as (high) priest (*see* priest(hood) of Jesus)
  as new/Second Adam 12, 15, 27, 32–5, 36, 39, 42–3, 45, 52, 55, 71, 94, 110, 125, 130
  as radiance/glory of God 16–19, 23, 26, 33
  as Son 5, 9, 10–12, 15–26, 27–8, 30–6, 37–9, 42, 45, 46–9, 51, 52–3, 55, 56–9, 60, 62, 63–4, 68–70, 73, 75–6, 77, 81, 82, 83, 87, 88, 90, 91, 92, 93, 94, 95, 97, 101, 102, 105, 110, 111, 122, 125, 130
John Chrysostom 71, 75
Joseph 63, 108, 110
Joshua 46–9, 51, 53, 56, 57, 106
Jubilee, Year of 47, 53, 83–4

law, Mosaic 74, 76, 79, 82–4, 85–6, 94, 124
  human 113
  natural 98
Levi, *see* priest(hood), Levitical

martyr(dom) 103–4, 114
Melchizedek 55, 64, 68–73, 76–7
Moses 20, 22, 37, 39–40, 46, 49, 57, 80–1, 95, 110, 116, 119–20
Mount Sinai 12, 49, 115, 119–20
Mount Zion 12, 115, 120–3

New Jerusalem 12, 85, 115, 119–23, 124
Noah 63, 108–10
  Noah (Darren Aronofsky film) 108

ocean 8, 36, 43, 118

plastic 8, 36, 43, 118
Plato(nic) 4–5
priest(hood) 5, 10, 35, 55, 57, 59–60, 67, 69, 71, 72, 74–5, 77, 80, 82, 84, 88–9
  of Aaron 57–8
  of Adam/human beings 11, 12, 29, 30, 34, 35, 37, 38–9, 41, 42, 50, 53–4, 55, 58, 61, 63, 67, 72, 73, 74, 75, 77, 81, 87, 89, 92, 96, 97, 99, 102, 108, 110, 117, 119, 125, 127, 130
  of Jesus 10, 27, 34, 35, 42, 55, 56–9, 62, 63, 64, 67–77, 79–80, 82, 87, 88, 89, 90, 92, 93–4, 125, 130
  Levitical 10, 55, 67, 68–9
  of Melchizedek 55, 58, 59, 62, 63, 68–70, 73–4, 75, 76, 93
prophet(ic) 7, 13, 15, 16, 20, 22, 26, 38, 39, 83, 86, 90, 103, 127

rest 40, 45–51, 52–3, 83, 89
revelation 5, 11, 15, 16, 18, 19, 20, 23, 26, 38

Sabbath 47, 48, 50, 52, 53, 83–4, 86, 87, 126
sacrifice 5, 57–8, 74, 79, 80, 87–7, 99, 107, 109, 111, 124–5, 126–7
sanctuary 5, 79, 81–2, 84–5, 87, 88–9, 91, 93–5, 122
shadow 48, 49, 80, 84–5, 87–90
sin(ners) 12, 19, 22, 41, 56–7, 58, 62, 74, 75, 80, 82, 89–97, 102, 108, 111, 116, 117, 118–19, 130
  cleansing/purification/atonement/ forgiveness of 10, 16, 18–19, 22, 25, 31, 34, 74, 82, 88, 89–97
spirit(ual)/spirituality 4, 5, 42–3, 47, 48–9, 50–1, 52, 55, 77, 106, 120, 125–6
subversion/subversive 9–11, 28, 35, 41, 45, 48, 49, 69, 70, 75, 91, 106, 130

tabernacle/temple 5, 38–9, 49–50, 51, 85, 87, 88–9, 93
Trinity/trinitarian 16, 25, 33, 36, 94

warning 15, 22–4, 26, 37, 39–42, 45, 51–2, 55, 60–1, 101, 102–4, 115, 116–19
wisdom 6, 25–6
  personified 25, 117

Ancient Document Index

Old Testament/ Hebrew Bible

Genesis
1:1–2:3	48, 49, 50–1, 121
1	19, 45, 51, 60, 67, 90
1:1	15
1:26-28	27, 28–9, 30, 34, 60, 74, 130
1:26	28
1:27	28
1:28	28, 34, 37, 62, 108
1:31	29, 48
2	45, 60, 67, 90
2:1-3	49
2:2-3	45
2:2	48, 52, 53
2:7	31, 33, 35, 59, 91
2:15	30, 60, 89, 130
2:19	31, 91
3	30, 41
3:8	51
3:17-19	30
3:17-18	61
6:7	90
6:17	91
7:15	91
7:22	91
8:21	109
8:22	109
9:1	63, 109
9:2	109
9:3	109
9:4	109
12:1	105
12:2-3	14
14	68
14:18-20	68, 69–70, 73, 75, 77
14:18	69, 70, 74
17:2	62, 108
17:6	62, 108
17:8	62, 108
22:16-18	62, 108
22:17	62
24:37	107
26:3-4	62, 108
26:24	62, 108
28:3-4	62–3, 108
35:11-12	62, 108
47:27	110
48:3-4	63, 108

Exodus
1:7	110
17:1-7	40
19	119
19:12	119–20
20:8	86
20:9-11	86
22:17	62
23:10-12	46–7
25–40	49
32	41

Leviticus
16	10
17:11	94
25–27	46
25–26	83
25	47, 53
25:1-12	83, 86
25:1-7	46
25:8-12	47
26:3-45	47
26:34-35	47, 83

Numbers
14	40
14:20-23	40
20:2-13	40

1 Samuel
15:22	90

2 Chronicles
36:20-21	47, 83

Job
38–41	97

Psalms
2:7	57
8	12, 30, 31, 32, 34, 36
8:3-4	28
8:5-7	28, 35, 60
8:6	28
8:7	28
39:6-8 LXX	91–2
39:6 LXX	92
40:7-9	91–2
40:7	91–2
94:7b-11 LXX	40–2
94:11 LXX	48, 53
95	42, 47, 52
95:7b-11	40–2
95:8	40
95:11	40, 46
101:26-28 LXX	20, 105
102	22
102:25-27	20–2, 105
103:4 LXX	21
104	22, 97
104:4	21
104:29	22

104:30	22	7:27	25	8:18-25	19, 21, 108, 112
104:35	22	7:28	25–6	12:2	13
110:4	58, 68, 82	8:1	25		
		11–19	117	1 Corinthians	
Proverbs		11:15-16	117	8:6	25
3:11-12	116	15:14-19	41	9:9-10	91
3:19	25				
8:22-31	25	**Pseudepigrapha**		2 Corinthians	
		1 Enoch		3:3	84
Isaiah		45:1	121		
1:10-13	90	*2 Baruch*		Colossians	
13:1-22	121	32:1	121	1:16-17	17, 25
61:1-2	53	59:3	121	1:16	11
		Jubilees		1:20	17
Jeremiah		1:29	121	2:9	73
7:21-24	90	*Sib. Or.*		3:5	118, 126
31:31-34	82–5, 88	3:675-80	121		
31:33-34	88			Hebrews	
		New Testament		1:1–2:4	11, 15–26
Daniel		Matthew		1	27–8, 38, 56
2	122	5:17	86	1:1-4	16–19, 23
2:31-35	122–3	5:20	86	1:1	38
2:36-45	122–3	6:26	90	1:2-3	11, 22, 25–6, 60, 125
2:44	122	8:23-27	24	1:2	12, 16–18, 33, 38, 51, 60, 70
2:45	122	Mark			
		4:35-41	24	1:3	16, 18–19, 22, 38
Hosea					
4:3	108–9	Luke			
6:6	90	4:16-21	53	1:4	20
		4:19	53	1:5-14	20–2
Amos		8:22-25	24	1:5	57
5:21-26	90			1:7	21, 23–4
		John		1:10-12	20, 105
Micah		1:3	25	1:11	20
6:6-8	90	1:10	25	1:12	22
		1:14	31	1:14	21, 24
Zephaniah				2	27–8, 32, 38, 56
1:2-3	90	Acts			
1:3	108–9	1:5	23	2:1-4	22–4, 26
		1:8	23	2:2	20, 38
Haggai		2:2	23	2:3	23
2:6	120–1	2:3	23	2:4	23, 39
				2:5–10:18	12
Apocrypha		Romans		2:5-18	12, 27–36, 60
Wisdom of Solomon		1:18-25	41	2:5-9	27–32
7:22–8:1	25–6	1:22-23	117	2:5	27–8
7:22	25	8:4	86	2:6-8	28, 34, 60
7:23	25	8:18-27	42		
7:26	25				

2:7	28	6:3	60	9:11–10:18	89–97, 102
2:8	30	6:4-6	60–1	9:11-12	91, 93
2:9	30–2	6:4	39	9:11	5, 122
2:10-18	32–5	6:7-8	61	9:12	91, 95
2:10	32, 38, 56, 60	6:9-12	61	9:13-14	74
2:11	34	6:13-20	62–4	9:13	91
2:12-13	34	6:13-14	107	9:14	39, 42, 94
2:14-15	60	6:14	62	9:18-22	93
2:14	33–4	7	55, 56, 58, 61,	9:19	91, 95
2:17	34, 68		64, 68, 73,	9:23-24	91, 93
3	37–8, 46, 47		76, 80, 85	9:23	90
3:1-19	37–43	7:1–10:18	59	9:24	90, 122
3:1-6	37–9, 81	7:1-28	67–77	10:1-10	91
3:2	38	7:1-10	68–73	10:1	90
3:4	38, 81	7:1	69	10:4	91, 95
3:5	38	7:2	69, 76	10:5	91–2
3:6	38–9, 40, 81	7:3	69	10:6	92
3:7–4:11	102	7:4-10	69	10:15	39
3:7-19	39–42	7:7	69	10:16-17	88
3:7	39, 40, 48, 52	7:11-28	70, 73–6	10:19–11:40	101–14
3:13	40, 48	7:12	74, 82	10:19-39	102–4
3:15	40, 48	7:13-14	73	10:19-25	102
3:16-18	41	7:16	74	10:19	102
3:17	41	7:17	68	10:26-39	102
4	38, 39, 40	7:18	82	10:29	39, 102
4:1-13	45–54	7:19	74, 82	10:32-39	103
4:1-11	45–51, 106	7:21	82	11	12, 51, 101,
4:3-4	48	7:26-28	74		104, 105,
4:3	47–8	8	74, 79–80, 81,		107, 110,
4:6	48		82, 84, 88		116
4:7	46	8:1-13	79–86	11:1-40	104–11
4:8	46	8:1-7	79–81	11:1	105
4:9	48	8:1	80	11:3	105, 110
4:10	48	8:2	80, 81	11:7	110
4:11	48	8:5-6	5	11:8-16	104–7
4:12-13	51–2	8:5	80	11:8	105, 110
4:12	51	8:6	82	11:9	105, 106,
4:14–6:20	55–65	8:7	82		110
4:14–5:10	56–9, 68	8:8-13	82–4	11:10	104, 105
4:14	56	8:8-12	88	11:13	106, 107
4:15	56	8:8	82	11:14-16	106
5	57, 59	8:10	94	11:16	104
5:3	74	8:13	20, 80, 82	11:17-19	111
5:4	57	9:1–10:18	87–99	11:21	110
5:5-6	58	9:1-10	88–9	11:22	110
5:6	58, 68	9:1-5	88–9	11:23	110
5:7	58	9:6–10:18	10	11:29	110
5:11–6:12	59–62	9:6-10	88	12	22, 51, 116
6	59	9:8	39	12:1–13:25	115–27
6:1	59	9:10	89	12:1-17	116–19

12:1	104, 116	1 Peter		*Panarion*	
12:2	116	1:1	106	4	70
12:3	116	2:5	74, 125		
12:4	116	2:9	74, 125	Eusebius of Caesarea	
12:5-11	116	2:11	106	*Proof of the Gospel*	
12:16-17	119			5:3	70
12:18-29	12, 105, 106, 119–23	2 Peter		Gregory of Nazianzus	
12:18	120	3:3-13	112	*Epistle 101*	31
12:20	119	Revelation			
12:22-24	120	1:6	74	Gregory of Nyssa	
12:22	120, 123	5:10	74	*On the Making of Man*	
12:25-29	120, 122	21–22	21		
12:26-27	12	21:1-4	85	2–5	30
12:27	12, 120, 121–2	21:22	51, 85	Jerome	
13	124	**Early Christian Writings**		*Hebrew Questions on Genesis*	
13:1-25	124–6	Clement of Alexandria			
13:8	124–5	*Stromateis*		14:18-19	70
13:9-16	124	4:25	70		
13:9-14	124, 125			Maximus the Confessor	
13:9	124	Cyprian		*Ad Thalassium*	
13:10	70–1, 124	*Letter*		51	29–30
13:11-13	124	62:4	70		
13:14	124			Photius	
13:15-16	124, 125	Epiphanus of Salamis		*Fragments on the Epistle to the Hebrews*	
13:15	125	*Against Melchizedekians*			
13:16	125–6	6:1-11	70	5:7-9	58
13:18-25	124				

Modern Authors Index

Andreopoulos, Andreas 71–2
Attridge, Harold W. 48, 95, 120

Badke, William B. 113
Baker, Sam 118
Barker, Margaret 49–50
Bartholomew I 64
Bauckham, Richard 29, 65, 98
Beale, Gregory K. 49
Bordeianu, Radu 30
Brown, William P. 21, 49
Buchanan, George W. 121
Butt, Nathalie 103
Byrne, Brendan 112

Cadwallader, Alan H. 104, 107
Campbell, Constantine R. 123
Carson, Rachel 2
Chryssavgis, John 26, 64
Conradie, Ernst M. 59–60, 64, 129–30
Craddock, Fred B. 48, 106
Crossan, John Dominic 127

Easter, Matthew C. 130
Edwards, Denis 71, 75
Edwards, Lucy E. 64
Ellingworth, Paul 48, 120

Fewster, Gregory P. 19
Foltz, Bruce V. 64
Francis (Pope) 64
Fretheim, Terence 29, 30, 110, 117

Goodall, Jane 118
Green, Chris E. W. 75–6

Habel, Norman 1–2, 4, 6–7, 29, 30, 46, 58, 64, 76, 106–7, 108, 116, 117, 131
Hatzidakis, Emmanuel 71
Hobgood-Oster, Laura 97
Hogue, Michael S. 97–8
Horrell, David G. 112

Huddleston, Jonathan 49
Huff, Peter A. 97
Hunt, Cherryl 112

Imhoff, Daniel 98

Jewett, Robert 48, 107
Johnson, Luke Timothy 48, 92, 95

Koester, Craig R. 48, 106

Laansma, Jon 19, 49, 121
Lambrick, Frances 103
Lamp, Jeffrey S. 1, 17, 24, 25, 29, 31, 39, 41, 43, 46, 48, 52, 53, 61, 64, 71, 90, 105, 106, 108, 117, 121, 124, 131
Lane, William L. 40, 48, 57–60, 81–3, 92, 106, 116, 121
Linzey, Andrew 97
Lodahl, Michael 97
Louth, Andrew 30
Lowe, Ben 65

McFague, Sally 65
Menton, Mary 103
Michaels, J. Ramsey 19
Middleton, Richard 29
Moltmann, Jürgen 64
Morgan, Jonathan 95–6
Munteanu, Daniel 26

Northcott, Michael S. 96

Painter, John 50
Peeler, Amy L. B. 18

Renwick, Anna 103
Rhoads, David 76
Richter, Sandra 30

Santmire, H. Paul 76
Scandrett, Joel 65
Schaeffer, Francis A. 29

Schenck, Kenneth L. 19–22, 48, 105, 120, 130
Schmemann, Alexander 72–3
Sleeth, J. Matthew 126
Snyder, Howard A. 65
Southgate, Christopher 112
Spencer, Nick 53
Stewart, Robert B. 127
Stone, Lawson 109
Swoboda, A. J. 24, 65, 126

Theokritoff, George 73
Tonstad, Sigve K. 19
Trainor, Michael 24

Van Dyke, Fred 47, 64–5, 83
Vroblesky, Virginia 53

Wade, Richard 98
Walton, John H. 8
White, Lynn, Jr. 6–7, 29–30
White, Robert 53
Williamson, Ronald 70
Wilson, Edward O. 2
Witherington, Ben, III 94, 121
Wright, N. T. 32, 49, 54, 62–3, 104, 108–10, 123, 127

Zizioulas, John 73

www.ingramcontent.com/pod-product-compliance
Lightning Source LLC
Chambersburg PA
CBHW070643300426
44111CB00013B/2233